The Evolution of Primate Behavior
Alison Jolly

Primate Evolution
Elwyn L. Simons

The Macmillan Series in Physical Anthropology
Elwyn L. Simons and David Pilbeam, editors
other volumes in preparation

David Pilbeam

YALE UNIVERSITY

THE ASCENT OF MAN

AN INTRODUCTION TO HUMAN EVOLUTION

The Macmillan Series in Physical Anthropology

THE MACMILLAN COMPANY, NEW YORK
COLLIER-MACMILLAN LIMITED, LONDON

The Macmillan Company
866 Third Avenue,
New York, New York 10022

Collier-Macmillan Canada, Ltd.,
Toronto, Ontario

Library of Congress catalog card number: 71-153767

PRINTING   3456789   YEAR   3456789

Preface

In the century since Charles Darwin wrote *The Descent of Man*, the physical evolution of man from some ape-like or monkey-like ancestors has become quite widely accepted as fact. For Darwin, evidence of man's affinity with other primates was based mostly on anatomical similarities. Thus, in the *Descent*, Darwin points out that

> It is notorious that man is constructed on the same general type or model with the other mammals. All the bones in his skeleton can be compared with the corresponding bones in a monkey, bat, or seal. So it is with his muscles, nerves, blood-vessels, and internal viscera. The brain, the most important of all the organs, follows the same law.

Since Darwin's hypothesis, the fossil evidence accumulated has, first, overwhelmingly attested that human evolution did occur and that man and the other primates are cousins, sharing common ancestors, and, second, allowed us at least an over-all view of the course of human evolution. Particularly since 1940, the amount of fossil material directly relevant to human and other primate origins has increased dramatically.

Reliable studies of living mammals under natural conditions—especially primates—have enriched our knowledge of the relationships between structure, function, behavior, and ecology. These new data have, in turn, enabled us to talk more realistically about function and behavior in extinct primates, including man's ancestors. Since natural selection favors adaptive behaviors rather than simply morphologies (although, obviously, the two are linked), we are now in the position to construct realistic dynamic models of human evolution.

The evolution of man's distinctive social behavior, particular social communication, and, above all, his culture, is to most of us still a mystery. It is this aspect of human evolution that has proven most difficult to grapple with, for it requires a detailed understanding of the social life of nonhuman primates and requires coordination of data from fields as diverse as ethology, neurophysiology, and comparative psychology. Potentially, this is the most exciting of all areas related to human origins. It is hoped this same book, revised ten years from now, will not only have the fossils sorted out properly but also propose models for social evolution that are plausible, perhaps even testable.

Any author clearly owes considerable debts to others. My thanks go to colleagues and students here at Yale—particularly to Elwyn Simons—for providing a stimulating yet relaxing atmosphere in which to work; to colleagues elsewhere, particularly F. Clark Howell and Alan Walker; to my editors at Macmillan, especially Charles Smith and Ray Schultz; finally, though, my thanks go to Carol, my wife, to whom this book is dedicated.

D. P.

Contents

Introduction 1

Modern man is one of the most successful mammals that ever lived, successful entirely because of the development of cultural behavior. "Culture" describes many of the aspects of behavior that are learned and transmitted from one generation to another, mainly through the medium of language. Political systems, myths, kinship relationships, incest taboos, rituals of various sorts, tool making, all are part of "culture." In Ralph Holloway's terminology (1969), culture is the imposition of arbitrary form upon the environment, and it is the combination of arbitrariness and imposition that makes human learned behavior so different from the undoubtedly learned, and complex, behavior of other primates. For example, only man can know the difference between fluoridated water and holy water (which might also be fluoridated), because only man can give arbitrary descriptions to objects, concepts, and feelings and communicate meaningfully these things to other human beings.

"Modern man" (*Homo sapiens sapiens* in zoological terms) became widely dispersed throughout the Old World only within the last

40,000 to 50,000 years. The word "only" is used because this represents but a small fraction of the total time since the hominids (man and his immediate and recognizable ancestors and relatives) diverged from the lines leading to the modern apes, or pongids. The split probably occurred at least 10 million years ago, and perhaps closer to 14 million or 15 million years ago (Simons and Pilbeam, 1965).

During this time, the hominids evolved their highly unusual and typical characteristics; the most important of these are complex cultural behavior dependent upon a relatively large and reorganized brain, a habitual upright bipedal locomotor pattern, and an unusual dentition in which the canine teeth are small in both sexes and the cheek teeth have large flat grinding surfaces (Le Gros Clark, 1964). Brain enlargement, associated with changes in the internal organization of the brain, began at least 3 million years ago (Holloway, 1970). By this time, too, bipedalism was well established. The dental changes can be dated to a much earlier time—more than 10 million years ago—and apparently indicate an important shift in the behavior of the earliest direct human ancestors from tree feeding to ground feeding. What follows is an attempt to outline what is known of the human fossil record, what happened during the course of man's evolution, and why it happened.

If we turn to the characteristics of man that are of most use to paleontologists (that is, the features that can be detected in the fossil record in one way or another), "modern man" is distinguishable from his immediate ancestors in a number of ways, principally relating to the structure of the skull (Le Gros Clark, 1964). The brain case of modern man is short from front to back, high, and relatively narrow, with an evenly rounded occipital region and more or less vertical frontal area which lacks heavy brow ridges. The face is relatively small and nonprojecting, and the dentition is, on the whole, smaller than that of earlier hominids. The brain of *H. s. sapiens* has an average volume of around 1,400 cm^3. Essentially, the bones of the brain case are molded closely to the external contours of the brain. To a considerable extent, the facial skeleton is tucked under the front of the brain case so that the skull has a rather globular appearance. The dentition is arranged in a continuous parabolic curve; both sexes have small canines that are chisel-like teeth similar to the incisors. Incisors and canines function together as slicing teeth, and the premolars and molars behind function as grinders. Man is a very efficient upright biped; the average stature of most living groups varies between 5 feet 3 inches and 5 feet 9 inches, a somewhat small range of variation.

Modern man today is a polytypic species, that is, a species divided into a number of populations that generally inhabit different geographical areas (Mayr, 1963). These populations do intergrade—they are not genetically isolated like species are—yet each population is generally homogeneous in certain traits. Thus average skin color, or head form, might vary from group to group. Some of these differences are thought to be adaptations to local environ-

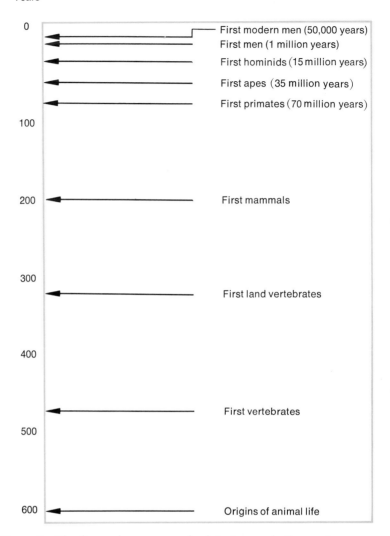

Millions of
Years

0 — First modern men (50,000 years)
— First men (1 million years)
— First hominids (15 million years)
— First apes (35 million years)
— First primates (70 million years)

100

200 — First mammals

300 — First land vertebrates

400

500 — First vertebrates

600 — Origins of animal life

Figure 1 The figure shows on an absolute time scale the most important evolutionary events leading to the appearance of modern man. (Pilbeam.)

mental conditions. Almost all widely ranging species are polytypic, showing geographical differentiation in certain traits, and we can assume that the hominids, as long as they were wide ranging, were similarly polytypic. This is important in evaluating the fossil evidence for human evolution, because we must take into account not only variation within a local population but that between populations, too, as well as that resulting from species evolution through time (Pilbeam and Simons, 1965).

Most modern men live in urban communities, are pastoralists, or are herders. Very few indeed, less than 0.001 per cent of a total of 3 billion human beings, live at the subsistence level known as hunting and gathering. Hunters and gatherers live in small groupings—around fifty or so individuals—and obtain food by foraging and by hunting. There is a marked division of labor between the sexes, the women gathering plant food while the men hunt. Little or no economic surplus is ever accumulated, and economic and other social interactions are almost exclusively reciprocal. Sharing and cooperation are characteristic and at a premium. The bands themselves consist of a number of rather loosely knit nuclear families—husband and wife plus their offspring—and may fragment or coalesce with other groups depending upon environmental conditions (Lee and DeVore, 1968).

The hunters of today live in marginal habitats—Bushmen in the Kalahari Desert, Australian Aborigines in the Western Australian Desert—and are surrounded by nonhunting peoples. Often their behavior patterns have been modified by contact with peoples having other types of economic systems. However, a study of the living hunters does reveal a surprising consistency in certain features of social organization, and it does seem probable that these are correlated with their mode of subsistence. Some 10,000 years ago, all human groups, in a world population of perhaps 10 million, were hunters and gatherers. (Hunting as a way of life can support only a low population density.) At that time, certain populations began to cultivate plants and to domesticate animals. Food surpluses could then be accumulated, permitting an expanded population living in settled communities. The patterns of social organization and interaction changed markedly. Gradually, at first, the world of the hunters contracted, and their best hunting lands were taken for agriculture. Now they are close to extinction, or at least their economic system is. Nowhere do we find hunters living in ideal hunting conditions, surrounded by other hunters. Yet hopefully we can generalize, at least to a certain extent, from today's hunters to those of the past (Service, 1966).

As far as it is possible to tell, hunting as a way of life first became established at least 3 million years ago. At a conservative estimate, this type of subsistence level, and its correlated social organization, was characteristic of the hominids for at least 20 per cent of the time since they became separated from the apes, and this figure may well be too low by a factor of at least two. Human groups have been agriculturalists for less than a third of 1 per cent of the time since hunting became a basic hominid behavioral adaptation. It is the hunting way of life that has really molded man, to the extent of making him a large-brained biped, a tool-making cooperative carnivore, and, perhaps most important, a linguistic creature (Washburn and Lancaster, 1968).

Before the dispersion of modern man, other types of men were widely distributed throughout the Old World. They are classified as "archaic man" or *H. sapiens* (but not *H. s. sapiens*—a different

Figure 2 Part of a band of thirty-two Bushmen who have packed their kit and are heading to another waterhole. All their personal possessions are being carried by the group. (Courtesy of R. Lee.)

subspecies) because they had large brains and postcranially were almost indistinguishable from modern man (Le Gros Clark, 1964).[1] Archaic man's stone tools (Bordes, 1968) indicate a somewhat less sophisticated technology, but this was probably due not only to differences in innate ability but also to cultural (learned behavioral) differences. (Newer techniques just had not been invented.) Many of these people buried their dead with considerable ritual,

[1]We shall henceforth use the term "modern man" to mean *Homo sapiens sapiens* (also called by some "modern *sapiens*") and the term "archaic man" to mean *Homo sapiens* to the exclusion of *Homo sapiens sapiens* (archaic man is also called by some "archaic *Homo sapiens*," "archaic *sapiens*," or "early *Homo sapiens*"). And we will use the term "man" alone to include both "archaic" and "modern" man.

The species is the basic unit in any classificatory scheme, but there are other, more inclusive levels in the hierarchy. For example, species are grouped together into genera, and genera into subfamilies. Classificatory units at all levels of the hierarchy are known as "taxa" (singular: "taxon"). Thus, the species is a taxon, as is an order. Below are listed some of the most commonly used taxonomic categories.

Category	Example
Subspecies	*Homo sapiens sapiens*
Species	*Homo sapiens*
Subgenus	
Genus	*Homo*
Subfamily	
Family	Hominidae
Superfamily	Hominoidea
Infraorder	Catarrhini
Suborder	Anthropoidea
Order	Primates

Subfamily, family, and superfamily names have standard endings. Thus subfamily names end in "inae" (Ponginae), family names in "idae" (Hominidae), and superfamily names in "oidea" (Hominoidea). Such categories can also be written colloquially: for example, pongine, hominid, and hominoid.

and we must assume that their individual behavior and social organization were as advanced in some areas as those of later men (Howell, 1965).

These archaic men differed from modern men mainly in the form of the skull, which was long, low, and broad, with a big face surmounted by a massive brow ridge. What caused the changes in skull form between archaic and modern men is unknown. Possibly the brain became changed internally, resulting in an altered shape that affected the morphology of the skull, although this hypothesis is perhaps less likely than the view that mechanical factors (for example, head balance) were the dominant influences. However, the problem remains to be resolved.

Archaic-man populations were distributed throughout most of the Old World, populations showing variation between geographical areas in a way analogous to the pattern of variation in any polytypic species. The best known (and most famous) of these local populations are the neandertals, who lived in Western Europe during the last ice age (Howell, 1957).[2] Almost certainly, some

[2]"Neandertal" or "neandertal man" is the common name for *Homo sapiens neanderthalensis*, in the past commonly called "Neanderthal," "Neanderthal man," and so on. This subspecies is considered to be archaic man, not modern man.

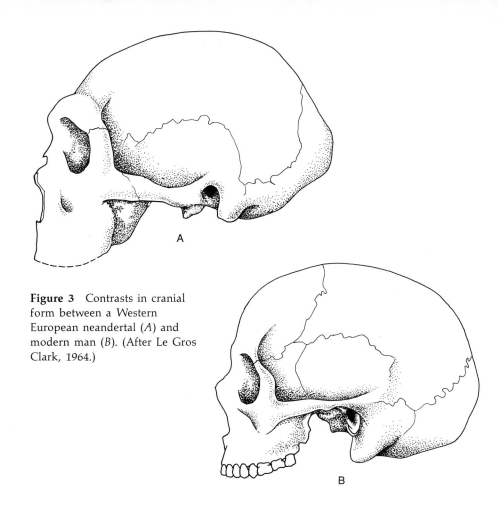

Figure 3 Contrasts in cranial form between a Western European neandertal (*A*) and modern man (*B*). (After Le Gros Clark, 1964.)

populations of archaic man, probably in East and North Africa and Western Asia, evolved into modern types. There is some reason to believe that local evolution from archaic to modern man occurred elsewhere in the world, too, although the evidence is regrettably sparse. Possibly these changes involved hybridization with more "modern" populations. However, in some areas archaic groups seem to have been replaced, apparently without very much interbreeding. Thus there is a mosaic of local evolution, migration, hybridization, and replacement in the transition from archaic to modern man.

Archaic man had appeared by at least 250,000 years ago. It is impossible to say exactly when the species evolved, because the boundary between this and the ancestral species is purely arbitrary, for we are dealing with a "lineage" (or evolving continuum), a continuous sequence of ancestral and descendant genetic species populations (Pilbeam and Simons, 1965). (A genetic species is the largest naturally occurring population that interbreeds and that is capable of producing fully fertile offspring. Obviously, it has little or no "time depth.") Because lineages evolve, as we pass backward in time from modern man we will begin to recognize populations that are progressively less like man. At some point, it will no longer be meaningful to describe the populations as *Homo sapiens,* so they are given another species name—*Homo erectus.* The word "species" in this case describes a set of individuals that has a time depth; it is a sequence of genetic species, a segment of a lineage. Lineages and genetic species are biological realities; that is, they are non-arbitrary as far as inclusion and exclusion are concerned. Reference of individuals to these sets depends solely on the criteria of inter-breeding, viability, and fertility. However, segments of lineages, or time-successive species as they are called (such as *H. sapiens*—in this sense—and *H. erectus*), are arbitrary sets. Their temporal boundaries are set entirely for convenience and are usually drawn at a time where few or no fossils are known; the absence of fossils creates the impression of a gap, which actually of course, did not exist. As more fossils become known, the gaps get filled in and the "boundaries" between time-successive species blur, eventually disappearing. Obviously, the latest *H. erectus* and earliest *H. sapiens* will resemble each other very closely and will differ markedly from both early *H. erectus* and late *H. sapiens.* This is no problem—or should not be. We need to divide lineages and to give the segments different names, or otherwise we should be describing Miocene apes as *H. sapiens.* But the segments themselves are arbitrary. The differences between *H. erectus* and *H. sapiens*—adjacent parts of one lineage—are in no way the same as the differences between, for example, *Pan troglodytes* (the chimpanzee) and *Gorilla gorilla,* related but distinct living species. *Homo erectus* is the name given to the evolving hominids who lived between about $1\frac{1}{4}$ million and $\frac{1}{2}$ million years ago (Howells, 1966). The species is characterized by a relatively small brain, averaging around 800 to 1,000 cm^3 for various populations, a distinctive cranial morphology, and face and dentition larger than those of most *H. sapiens.* Postcranially, as far as is known, *H. erectus* was indistinguishable from *H. sapiens* in terms of locomotor efficiency and stature. The stone tools associated with *H. erectus* are in general (but only "in general") less diverse and less technologically sophisticated than those made by later hom-inids. However, some *erectus* populations used fire, and they were very successful big-game hunters, although perhaps not as effi-cient as the early *sapiens* groups. Most anthropologists believe that *H. erectus* was ancestral to *H. sapiens,* at least in a broad sense, and that the two "species" are similar enough to be placed in one genus, *Homo.*

The evolution of *H. sapiens* from its antecedents occurred in just the same manner as the transition from archaic to modern man, by a combination of local evolution, population mixing, and re-placement. The ancestral–descendant relationship of *H. erectus* and *H. sapiens* has been questioned by a number of workers (Leakey, 1966), some suggesting that *erectus* was a side branch of the hominid line and that the ancestors of *sapiens* during the middle Pleistocene

Biologists use the term "species" to refer to different types of groupings. Thus a "biological species" can be defined as "the largest naturally occurring group of individuals which actually or potentially are capable of interbreeding with the production of fully fertile offspring."

Paleontologists use the term in another context. The definition of a "paleospecies" or "time-successive species" is "a segment of a lineage (an ancestral–descendant sequence of biological species populations) evolving through time, the time limits of which are arbitrary."

The diagram below represents a lineage evolving (changing genetically) through time. A and B are biological species that existed at times T_1 and T_2. X, Y, and Z are paleospecies, the temporal boundaries of which are arbitrary.

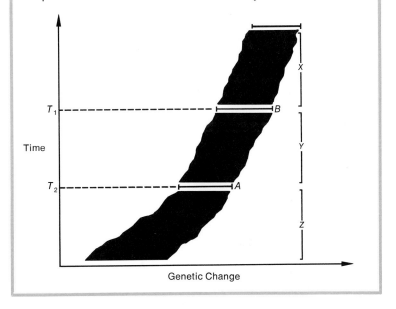

(the period between approximately $1\frac{1}{4}$ million to 1 million and about $\frac{1}{2}$ million years ago) are unknown. As we shall discuss later, this argument is rather implausible.

The ancestral hominids of the late Pliocene and early Pleistocene (roughly the time from 5 million to 1 million years ago) are generally placed in the genus *Australopithecus,* separate from *Homo* (Le Gros Clark, 1964). It is probable that two successive species were ancestral to *H. erectus,* the youngest, *A. habilis,* going back to around 2 million years ago (Leakey, Tobias, and Napier, 1964) and the older, *A. africanus,* to some 5 million years ago. These hominids were basically very similar, although *A. habilis* had a larger brain (600 cm³ on average compared with just over 400 cm³) and somewhat smaller teeth than *A. africanus.* The dentition in both species—particularly the cheek teeth—was both absolutely and relatively (taking account of body size) larger than in later hominids. Canines and incisors were human in shape and orientation, implying similar functions. Both were small bipeds, around 4 feet tall, weighing some 50 pounds on average; there is evidence to suggest that their locomotion was less efficient than that of *Homo* species.

Stone tool kits are known now from deposits as old as $2\frac{1}{2}$ million to $2\frac{3}{4}$ million years (Howell, personal communication), and it seems quite possible that stone tool making is more ancient still. The existence of these stone tools implies that the hominids who made them had become hunters, and it is clear that these hominids had developed home bases and had ceased to be nomadic like other primates. Evidence from animal remains associated with *Australopithecus* supports this conclusion, as does the habitat—open woodland and savanna—they occupied. Recent studies of human and ape brains, and of brain casts from these early hominids, suggest that the brain of *Australopithecus* had become reorganized in certain specifically human ways (Holloway, personal communication).

Figure 4   Casts of Oldowan stone tools (below) and actual Acheulian hand axes. The Oldowan Industry first appeared in Africa a little over $2\frac{1}{2}$ million years ago. The earliest hand axes were made a little over 1 million years ago. (Pilbeam.)

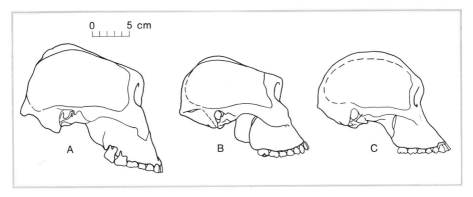

Figure 5 Side views of the crania of three *Australopithecus* species: *A. boisei* (A), *A. robustus* (B), and *A. africanus* (C). (After P. V. Tobias, 1967.)

Add to this the fact that these species were upright bipeds and we can reconstruct their behavior as being much more similar to that of man than that of the apes. They were hunters and probably lived in groups characterized by division of labor, strong ties between male and female, and strong and long-lasting mother–infant ties. The presence of tools, the more refined hand–eye coordination necessary for their manufacture and use, the shift to bipedalism, and the changes in external brain morphology all point to a reorganization of the hominid brain to one that was capable of generating and mediating specifically human behavior patterns such as language. Tool making and language are similar operations; they impose an arbitrary form on the external world (for example the concept "arbitrary" could be conveyed by any combination of sounds). They involve sequential operations, these sequences being "nested" or hierarchical. In terms of output, an animal with a brain capable of generating tool-making behavior would most likely be capable of generating some kind of language.

So, our ancestors were open-country hunters for at least 3 million and perhaps as long as 4 million or 5 million years. Living alongside the two *Australopithecus* species discussed above were other primitive hominids, members of at least one other lineage. This lineage is known generally as *A. boisei* (Tobias, 1967), although possibly a different generic name is justified. These hominids were larger than either *africanus* or *habilis,* ranging up to perhaps 150 pounds or more in weight, and had enormous cheek teeth set behind diminutive incisors and canines. It is likely that *A. boisei* was adapted to a much more herbivorous diet than the other *Australopithecus* species and that it exploited a different ecological niche (that is, utilized different parts of the same habitat). The two lineages coexisted for some 2 million years, the robust *Australopithecus* eventually becoming extinct at the end of the early Pleistocene. The cause of the extinction is unknown. Perhaps as our ancestors became more successful, as their ecological niche expanded, our specialized cousins were outcompeted. Perhaps they were simply killed off. So, at least by the end of the Pliocene (around 2 million years ago), our ancestors were open-country hunters, tool-making, symboling bipeds, with already reorganized and enlarging brains. Little is known of the hominids of between 5 and 10 million years ago.

Ramapithecus (Simons, 1964) is a possible early hominid from the late Miocene of Africa and India and the early Pliocene of India (10 million to 14 million years ago). Unlike later hominids, it was a forest or forest-fringe form. It is known only from jaws and teeth, but what is known suggests that it was a ground-feeding herbivore, eating small vegetable food items. The canines of *Ramapithecus,* were reduced in size compared with those of apes, at least of females, and in the later forms were altered in morphology, too. The postcranial anatomy is unknown.

It has been argued for some time that the changes in dentition came about because the large male canine of primates was replaced functionally in display behavior and fighting by hand-held weapons (Washburn, 1960). No stone tools have been recovered with *Ramapithecus,* and it is perhaps unlikely that any ever will be. Possibly, objects were used as tools, just as chimpanzees today use twigs when fishing for termites or stones to crack open nuts. However, it is unlikely that these earliest hominids were continuously dependent on manufactured tools for survival, as we are, and as at least some species of *Australopithecus* were. It is difficult to reconcile the reduced canines of *Ramapithecus* with the (highly probable) absence of tools, if the relationship between canine reduction and tool use is correct. Rather, it seems more likely that the canines became reduced because of reasons involving masticatory function. Their reduction permits a considerable transverse shearing component in chewing that can be distributed along the entire cheek-tooth row. Also, once reduced they alter in morphology, becoming functionally incorporated with the incisors in a slicing battery set at the front of the face. The exact diet of the early hominids is unknown, but it apparently required a combination of slicing and grinding.

Thus the earliest hominids were probably vegetarian ground feeders, foraging in open areas around lakes or at the forest fringe. They were quite likely not habitual bipeds, nor does it seem necessary to assume that they were behaviorally advanced relative to the apes. Sometime in the Pliocene, as forests became replaced by open woodland with the general fall in temperature and increased seasonality of rainfall, hominids shifted from being vegetarian foragers to hunters and gatherers. This fundamental, and quite possibly rapid, behavioral shift eventually brought with it many other profound changes. The basic economic unit became a male plus a female, the male providing meat (and protection) for the less mobile female, who became increasingly tied to her infants, as their period of dependency increased with slower maturation.

Tool making was established, and quite probably the rudiments of vocal language, too. From this point on, hominids were cultural animals, imposing arbitrariness on the environment, thereby mak-

Figure 6 Tentative evolutionary history of the hominids. Solid lines represent known fossils, dotted lines represent probable relationships. (Pilbeam.) See note on page 156.

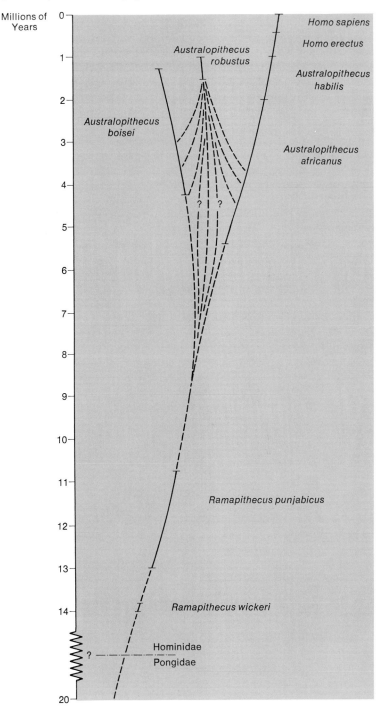

ing it more complex, increasing the richness of sensory input, and further selecting brains that were more effective processing organs. The slow process of reorganization and expansion culminated finally in the sophisticated and relatively enormous brain of *H. sapiens.*

Australopithecus species were bipeds, and probably quite effective ones, too (Le Gros Clark, 1964). Bipedalism became still more efficient in species of *Homo,* probably as a response to selective pressures associated with increasing home range, for both the hunters and the gatherers. These late Pliocene and early Pleistocene hominids also had home bases, camps where a band might spend several days or weeks, places where mothers and infants could remain while males were hunting, where meat could be butchered and shared and food stored. The home base was a place where sick or injured hominids might rest and be cared for; a broken ankle, or yellow fever, need not mean death, for a sick animal would no longer be denied the protection of the troop.

As hominid evolution continued through the Pleistocene, body size increased, tools became more complex, and brains got larger, probably as further refinements for hunting; these and other pieces of evidence point to a gradual increase in the complexity of cultural behavior, a gradual widening of the "cultural niche" until, by the middle Pleistocene, only one species of hominid was left, distributed throughout the Old World.

This chapter has outlined very briefly the course of hominid evolution, as far as we know it today. We shall develop some of the points made here in greater detail later, but first let us take a brief look at our closest living relatives, the primates.

Primates

2

Why study primates? For two principal reasons. First, they provide us with living models for analyzing the meaning of fossil remains; if we can relate, in extant species, structure to function and function to behavior and ecology, then it becomes potentially possible to say something about function in extinct forms. Second, the study of social behavior and social organization in primates can give us clues about the possible nature of the behavior and organization of early hominids, and, combining this with what we know of modern hunters and a knowledge of the archeological record, we can say something about the sorts of changes that have occurred, and why they happened.

PROSIMIANS

Primates (Napier and Napier, 1967) are an "order" of the class Mammalia and are divided, in turn, into two suborders, Prosimii (lower primates) and Anthropoidea (higher primates). The former are the relatively little changed descendants of the earliest primates (this we know from the fossil record), creatures that lived in North America, Eurasia, and probably Africa, too, between 70 million and 40 million years ago (Simons, 1961a, 1962). The early prosimians seem to have been mainly frugivorous (fruit eaters), unlike their insectivorous ancestors, and had relatively large brains compared with those of other contemporary mammals. They were arboreal creatures and evolved binocular vision (possibly color vision, too); early on, they developed the ability to climb trees by grasping. This was an important step, for it endowed them and later primates with dextrous hands and feet equipped with mobile thumb and big toe. These prosimians moved in a specialized way. Their mode of loco- motion is described as vertical clinging and leaping, and utilized predominantly vertical supports (Napier and Walker, 1967). The body was maintained at rest in a vertical position, powerful grasping feet supporting most of the body's weight while flexed arms pre- vented the trunk from toppling backward. Leaping involved only the long and powerful hindlimbs, which contacted the new support on landing. Living vertical clingers are adapted in numerous ways to this way of moving, and comparison of skeletons of the extinct prosimians with those of the living is what enables us to draw some conclusions about locomotor behavior in the earliest primates.

Order: Primates

Suborder: Prosimii
    Infraorder: Lemuriformes — Lemurs
        Indris
        Sifakas
        Lorisiformes — Lorises
        Bush babies
        Tarsiiformes — Tarsiers
Suborder: Anthropoidea
    Infraorder: Platyrrhini — New World monkeys
        Catarrhini — Old World monkeys, apes, and men
    Superfamily: Cercopithecoidea — Old World monkeys
        Hominoidea — Apes and men
    Family: Hylobatidae — Lesser apes
        Pongidae — Great apes
        Hominidae — Men

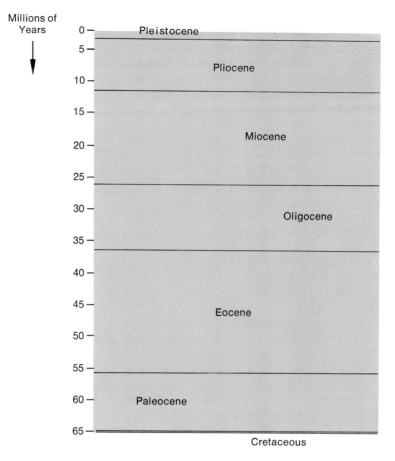

Figure 7   The Cenozoic Era, the time following the Cretaceous, is
subdivided into epochs as shown in the chart. The primates evolved
late in the Cretaceous, the earliest apes in the Oligocene, and the first
human ancestors late in the Miocene. (Pilbeam.)

The brains of these early primates were large, particularly in the
regions of the occipital and temporal lobes (Radinsky, 1967). These
areas are concerned, respectively, with perception and integration
of visual stimuli and with perception of auditory stimuli and storage
of visual and auditory memory. The same regions are similarly
developed in living prosimians, which strongly suggests that at least
some of the extinct species lived in groups in which the visual and
auditory channels were important in social communication. There
is also evidence to suggest that many of them were diurnal forms.

The number of prosimian species has become considerably
smaller since they first evolved. They are now confined either to
isolated areas (Madagascar) or to nocturnal niches. It is thought
that their numbers became reduced with the evolution of the higher
primates, which outcompeted the prosimians. The Madagascan

forms are particularly interesting. The behavior of some of these species has recently been studied by Dr. Alison Jolly (1966a), who found that a number of diurnal species live in multimale permanent social groups. This was a somewhat surprising discovery because it had been thought previously that such typical primate groupings could only exist in species with at least a "monkey level" of intelligence. The prosimians do not perform as well as higher primates on standard intelligence tests involving object manipulation and so forth. It now seems more likely than not that the typical primate type of social organization came first and was preadaptive to the evolution of higher primate-type intelligence (Jolly, 1966b).

Figure 8 *Propithecus verreauxi,* a Malagasy vertical clinger and leaper. (Courtesy of J. Buettner-Janusch.)

Figure 9 *Lemur catta,* a Malagasy quadrupedal lemur. (Courtesy of J. Buettner-Janusch.)

What gave the impetus to this evolutionary change is unknown. It has been suggested that a feedback relationship exists between the intelligence of predator and prey, that the evolution of more efficient and intelligent predators selects for the evolution of more successful and intelligent herbivores. Certainly, by Oligocene times (approximately 25 million to 35 million years ago) several groups of higher primates had evolved. These groups were characterized by new types of locomotor behavior, suggesting shifts in ecological niches, and by the evolution of a larger brain with greatly expanded association areas. The association areas are involved in the integration and elaboration of sensory input and motor output, among other activities.

NEW WORLD MONKEYS

One group of higher primates evolved in South America, and they are classified into an infraorder, the Platyrrhini (or New World monkeys) of the suborder Anthropoidea. One of these forms, the spider monkey of the genus *Ateles,* is of particular interest to us. It is a frugivorous form, feeding on ripe fruits, and lives in social groups consisting of twenty or more individuals. Because ripe fruits are, at any one time, widely dispersed, for foraging purposes the troop breaks up into a number of subgroups of varying composition. These units may consist of mothers with infants, males and females, or groups of males. The units aggregate or disperse depending on the availability of food but remain in communication

Figure 10 The spider monkey, *Ateles,* in typical hanging and feeding position. (Courtesy of D. Chivers.)

by means of loud vocalizations when spread out (Klein, personal communication).

The spider monkey is a terminal-branch feeder, moving in the outermost parts of trees among the smallest branches, and has evolved a number of quite specific adaptations to this mode of feeding. It has a prehensile tail, which it utilizes to support its body weight while suspending itself below branches (Erikson, 1963). A terminal-branch feeder cannot, unless it is very small, move on the tops of such small supports and therefore hangs beneath them. The

Figure 11 (opposite)   Skeletal contrasts between a quadruped (*A*, a capuchin), two arm swingers (*B*, a spider monkey, and *C*, a gibbon), and a knuckle walker (*D*, a chimpanzee). The quadruped has, in contrast to the others, a deep and narrow thorax and a long lumbar region. The chimpanzee, a knuckle walker, has an even more reduced lumbar region than the arm-swinging forms. *E* and *F* show the shape of the thorax viewed from above in a macaque (*E*) and man (*F*). Man is a typical hominoid in that the thorax is shallow and broad, the vertebral column is shifted relatively anteriorly, the clavicles are long, and the scapulae are set on the back of the thorax rather than on the side. (*A–D*: After Erikson, 1963. *E* and *F*: After Campbell, 1966.)

THE ASCENT OF MAN

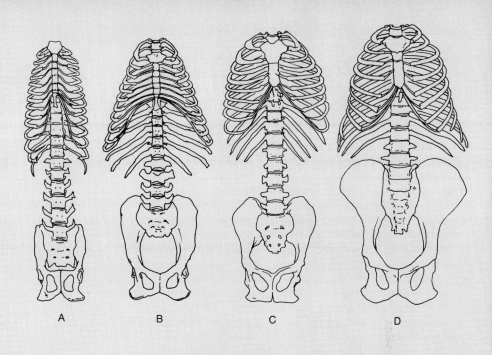

A   B   C   D

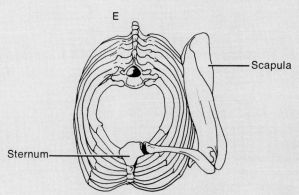

E

Scapula

Sternum

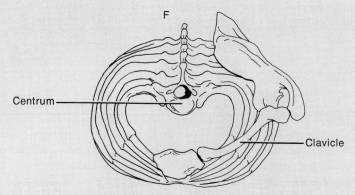

F

Centrum

Clavicle

spider monkey also hangs by its arms and can progress beneath supports using a combination of arms and tail, or arms alone. Such locomotion can be termed arm swinging, and inasmuch as the spider monkey is also quadrupedal (often using all four limbs while moving or at rest) we can describe this species as an arm-swinging quadruped. ("Brachiation" is another word for arm swinging, but it is preferable to restrict the use of this term to describe the locomotion of the small apes, the gibbons.)

A number of other New World primates also exhibit varying degrees of arm swinging and hanging in their locomotor repertoire, although none as much as the spider monkey. *Ateles* has a number of adaptations to arm swinging, resulting from the facts that the trunk is often erect, or orthograde, and that the shoulder girdle is subject to tensile rather than compressive stresses during such activities. Also, the vertebral column serves as a short and relatively immobile strut during locomotion, unlike the long flexible backbone of strict quadrupeds, where springy flexion and extension of the vertebrae are important during running and leaping (Erikson, 1963).

Thus, the forelimbs are long relative to trunk length, this increasing "stride length" during arm swinging. The wrist, elbow, and shoulder joints are remodeled, providing a combination of stability and mobility. The scapula (shoulder blade) has altered in shape (Oxnard, 1969a), principally to allow arm raising and rotation to be effected more easily and to cope with tensile stresses due to supporting body weight beneath branches. The hand is elongated and hooklike, with curved fingers and rudimentary thumb. The lumbar region of the backbone is short, and the joints in the column are so arranged as to permit less flexion and extension. The chest has broadened and become shallow from front to back, the scapulae moving to the back of the thorax rather than being on the sides as in orthodox quadrupeds. At the same time, the vertebral column has shifted forward, so to speak, thus bringing it closer to the center of gravity in orthograde (upright) posture. Hindlimbs are less well adapted to leaping than in other quadrupeds. This complex has arisen, it seems, as an adaptation to frugivorous feeding in a small-branch niche (Ellefson, 1968) and is of particular interest to us here because of the parallels to be drawn with the earliest ape and human ancestors.

OLD WORLD MONKEYS AND APES

Another group of higher primates evolved in Africa, again apparently during the Oligocene. These are classified into an infraorder, Catarrhini, and are further subdivided into two related but distinct groups, the superfamilies Cercopithecoidea (Old World monkeys) and Hominoidea (apes and man). The cercopithecoids and hominoids split at least 35 million years ago, perhaps more, as far as we can tell from the fossil record.

Very little is known about the earliest monkeys. For example, fewer than thirty are known from the Miocene (25 million to 35 million or so years ago) compared with many hundreds of hominoids from the same age (Simons, 1969a). This may be because Miocene sites are samples of ecological niches not inhabited to any great extent by monkeys. More likely, cercopithecoids were simply less diversified than hominoids at that time. It has been suggested by Dr. Clifford Jolly (1967) that the earliest Old World monkeys were specialized leaf eaters. One subfamily of living cercopithecoids, the Colobinae, is thought to represent the little-changed descendants of these early forms. For the most part, colobines are highly arboreal monkeys, feeding on a diet consisting almost entirely of leaves. They have specialized cheek teeth with high pointed cusps and sharp interconnecting crests as an adaptation to this diet, along with complexly sacculated stomachs in which the food is digested in ruminant-like fashion. They are basically arboreal quadrupeds; a few of them (for example, the *Nasalis* monkeys) do arm swing occasionally, although not nearly so much as does the spider monkey (Napier and Napier, 1967).

The second subfamily, Cercopithecinae, apparently evolved during the Miocene and became highly diversified and successful during the Pliocene. Cercopithecines are frugivorous and omnivorous creatures, although the larger ones also take small game, and have evolved to fill a wide variety of forest, woodland, and savanna niches. They live in various types of social groups. For example, the forest species of genus *Cercopithecus* apparently move in units containing only one adult male together with females and young (Struhsaker, 1969). Some of the most terrestrial of the monkeys live in similar groups, for instance, the patas monkey (Hall, 1965a) and gelada baboon (Crook, 1966). Other forest species, such as those of the genus *Cercocebus*, and more open-country forms, such as the vervets (*Cercopithecus aethiops*), the common baboons (*Papio cynocephalus*), and the macaques (genus *Macaca*), habitually move in multimale troops (Hall, 1965b).

The best-studied members of this group are the baboons and macaques. These are usually regarded as open-country animals, but as Dr. Clifford Jolly (personal communication) has recently pointed out, the treeless savannas, at least in Africa, are probably of very recent origin indeed and exist because the agricultural activities of recent man turned woodland savanna, woodland, and even forest into open grassland. Essentially, Jolly, along with other workers such as Dr. Thelma Rowell (1966), believes that baboons and macaques are basically woodland animals that have moved relatively recently into the developing savanna. Their social behavior studied in such habitats may therefore not be "typical" of that of the species (in fact, it should be remembered that species behavior is generally flexible enough to adapt and change in response to a variety of habitats).

Baboons in woodland live in multimale groups of rather flexible composition, according to Dr. Rowell (1966). Almost all males, and many of the subadult females, change troops during their lifetime. The integrity of a troop is maintained by the adult females with their offspring. Aggressive interactions between individuals are not as frequent as in baboons of the same species living in open country, and dominance hierarchies are less rigid. Savanna baboons, studied by a number of workers including Washburn and DeVore (1961a) and Hall and DeVore (1965), live in much tighter groupings, with little interchange of individuals between troops. A small number of adult males comprise a "central hierarchy," within which animals may form a more or less rigid dominance hierarchy, although this is infrequently linear. For example, baboon A may often be "dominant" to B (that is, be able to win fights, obtain access to desired food or females, and so forth), yet be "dominated" on certain occasions by a coalition of B and C.

Baboons, like many other cercopithecoids, are highly sexually dimorphic, the males differing from the females in body size, canine-tooth size, and aggressiveness. The earlier studies on savanna baboons suggested that the aggressiveness of the large males, coupled with their huge canines, was related to their function as defenders of the troop—which they certainly are in a savanna habitat. However, Rowell (1966) and others note that woodland and forest cercopithecines, exhibiting in many cases fully as much dimorphism as the grassland species, do not show the same protective male behavior against predators. Crook and Gartlan (1966) and others have suggested that the extreme sexual dimorphism is a result of intermale competition within the troop, an example of what Darwin called sexual selection. The large canines and general aggressiveness that the males evolved in this context are preadaptive to use in defense of the troop once a species moves into open country where ground predators pose greater dangers.

This is particularly important in any discussion of the evolution of the small canines of hominids, for their reduction need not imply that weapons were being used in their place against predators. Rather, it might imply that subtle and as yet poorly understood social behavioral changes had occurred, particularly as regards intermale relations, perhaps involving a shift away from aggressive toward cooperative behavior. Also, as already mentioned, there are some good functional reasons for canines to be small when the functioning of the hominid chewing mechanism is considered.

One area in which all monkeys are more advanced than prosimians is in the matter of infant care. A newborn baboon is more or less dependent upon its mother for the first 2 years of life, and ties between mother and offspring may last for the rest of their lives. Careful work by various Japanese primatologists on the Japanese macaque has shown that these ties exert a profound influence on the course of a monkey's life (Imanishi, 1960). For example, in the case of female offspring, the eventual dominance status of a daughter is closely tied to that of her mother, and experimental

Figure 12 Male and female hamadryas baboons, *Papio hamadryas*. Note the greater body size of the male, the longer face, and the well-developed mane. (San Diego Zoo Photo.)

work with laboratory colonies has shown that a daughter's status fluctuates with that of her mother if this is altered artificially. Sons are less influenced by their mother's status, because as adolescents they tend to associate far more with other older animals, particularly males, than do daughters, who generally remain close to their mothers. Siblings also tend to maintain close ties. Dr. Donald Sade (1965), who has studied macaques on Cayo Santiago in Puerto Rico, found that males who consistently spend time together frequently are "half-brothers." Thus the primate social group consists essen-

Figure 13 Mother–infant bonds are extremely close among primates. Here an infant baboon is shown clinging to its mother's belly as she moves to a new feeding place. (Courtesy of A. Walker.)

tially of matrifocal family groups, held together over long periods of time by the equivalent of kinship ties. The troop is centered around adult females; adult males, important for protection and reproductive purposes, are nevertheless frequently transients.

Sade (1968) and the Japanese workers have noted that male offspring almost never mate with their mothers; that is, there are biologically determined incest prohibitions. The occasional matings that occur do so only when sons become dominant to their mothers. However, this cannot be the only factor involved, since matings occur very infrequently even when sons do become dominant, and males often mate with other more dominant females. Apparently, certain aspects of the mother–son relationship produce psychological changes that actively inhibit mating.

The monkeys are phylogenetically much less closely related to man than are the apes. Some of them have nevertheless been used as models for the behavior of the earliest hominids when our ancestors first left the forests (Washburn and DeVore, 1961b). It is quite likely that they provide fewer clues to the social organization and behavior of early hominids than do the apes. However, they can be of some use in more specific ways, particularly when we come to examine the mechanical changes in jaws and teeth that occurred in the hominid lineage. However, before turning to that we should briefly outline some features of hominoid behavior and evolution.

The hominoids are subdivided into three families, the lesser apes or Hylobatidae (gibbons and siamangs), the so-called great apes or Pongidae (gorilla, chimpanzee, and orangutan), and the Hominidae. Studies of comparative anatomy (Schultz, 1968) and biochemistry (Sarich and Wilson, 1968) underline the interrelatedness of these forms and their distinctiveness from all other primates. All have brains relatively larger than those of monkeys, the expansion involving mainly the association areas of the cortex. Infants are also dependent on their mothers for a longer period, and the combination of these two factors means that more of the total behavior of an individual is learned and is therefore potentially more adaptable.

Hominoids also have common dental and postcranial characteristics, probably because they evolved originally from small, arboreal, frugivorous forms that exploited the small-branch niche in the forest canopy in much the same way as do the spider monkey and its allies among the New World monkeys (Napier and Napier, 1967; Jolly, 1967); the early hominoids seem then to have been quadrupedal arm swingers.

The apes and men have relatively long arms, short trunks with reduced lumbar regions, shallow broad thoraxes, scapulae on the back of the chest, and a whole complex of muscular adaptations associated originally with arm raising and hanging below supports. In adult gorillas these behaviors are very uncommon, and of course man is a biped, yet these features suggest at least a period of arm swinging during the ancestry of these forms. Differences between man and other hominoids are particularly marked in the shoulder girdle (Oxnard, 1969a,b), although it appears that these morphological changes have come about only during the past few million years.

THE GIBBONS AND SIAMANGS

The hylobatids—gibbons and siamangs—are the smallest of the apes, the largest weighing no more than 25 pounds or so. They live in dense tropical forest in Southeast Asia and have been distinct from the other hominoids for a considerable period of time, perhaps 30 or more million years, if the fossil record has been interpreted correctly. The living species are arboreal frugivores, characterized by a number of unusual adaptations (Ellefson, 1968; Schultz, 1968). They are the most acrobatic of all the primates; their locomotor pattern is described as brachiation, arm-swinging movement that contains a leaping phase when the body is propelled through space using the arms alone. This behavior occurs only in hylobatids. Brachiation and arm swinging may occupy up to 75 per cent of the time spent moving, the rest being spent quadrupedally or bipedally. Hylobatids almost never come to the ground.

As adaptations to brachiation, the hylobatids show all the hominoid adaptations of trunk and forelimbs; the arms are especially elongated. Gibbons and siamangs show almost no sexual dimorphism in body size or canine size, a feature very rarely found elsewhere among primates. An adult male and female live in a monogamous unit that apparently lasts for several years. This pair, plus young, lives in a tiny territory, about $\frac{1}{4}$ square mile in area, that contains preferred feeding trees. The territorial boundaries are quite rigidly defined and defended by the male (unusual behavior in higher primates, where a troop generally moves through a home range that overlaps with the ranges of other groups). Males of adjacent groups fight frequently over territorial boundaries.

Dr. John Ellefson (1968), who has recently studied the common white-handed gibbon, *Hylobates lar,* has concluded that brachiation is an inefficient method of moving over long distances; he views it rather as a feeding adaptation. It is therefore possible to argue that as hylobatids became ever more efficient brachiators the feeding area through which they could effectively move became more restricted, and this in turn was associated with a reduction in group size as well as territory size. The smallest number of adult members of a social group is two, one male and one female. Once the group size is so reduced, it might well have been adaptive for the female to resemble the male in body size and canine size; perhaps the female acts occasionally in aggression and display as an "additional male." Thus it is possible that brachiation, territoriality, pair bonds, and lack of sexual dimorphism are combined in an adaptive package in hylobatids. (Territoriality, pair bonding, and reduced dimorphism are also found together in some New World monkey species [Mason, 1968].)

What about hylobatid ancestry? Gibbons were distributed more widely in China and Southeast Asia earlier in the Pleistocene (Hooijer, 1960) and one gibbon-like tooth has been found in early Pliocene deposits (around 11 million years old) in India (Simons, personal communication). Species of a primate with rather gibbon-like skulls, jaws, and teeth are quite plentiful in European sediments ranging in age from middle Miocene (16 million years) to early Pliocene (10 million years). These are classified into the genus *Pliopithecus* (Hürzeler, 1954). Earlier species of similar type, classified as *Limnopithecus,* are known in Africa from deposits in Kenya and Uganda and range in age from around 23 million to 14 million years (Le Gros Clark and Thomas, 1951). The two genera are closely related and are best placed in a subfamily, the Pliopithecinae, that probably belongs in the family Hylobatidae, although, as we shall see, this is by no means certain.

Pliopithecines have been regarded as early hylobatids by a number of workers (Zapfe, 1958; Simons, 1960), principally because of resemblances in jaws, teeth, and skulls. However, we are now

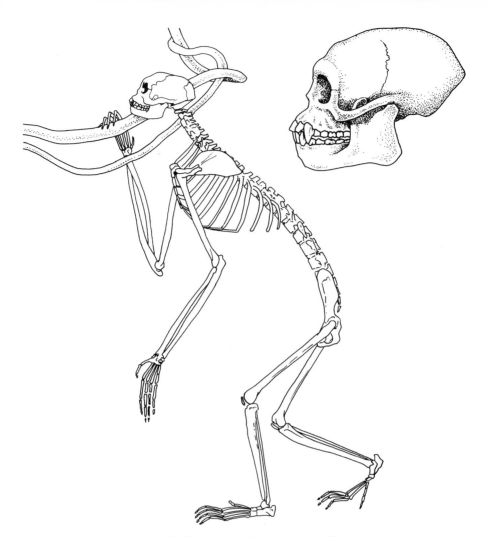

Figure 15 Reconstruction of *Pliopithecus,* a late Miocene gibbon relative from Europe. Although still possessing many quadrupedal traits, *Pliopithecus* was probably evolving a locomotor pattern involving arm swinging. (After Simons, 1964a.)

beginning to go beyond simple resemblances in asking questions of paleontological data, for it is becoming generally realized that a structure is the way it is for good functional reasons. Thus, gibbon teeth are adapted principally for fruit eating and have low, rounded cusps, united by rather blunt ridges. If the pliopithecines were also small frugivorous hominoids, would not their teeth be similar, even though they might not be ancestral to living hylobatids? This question cannot really be given a satisfactory answer now, because it requires very detailed analysis of the relationships between structure and function in various living species (including those frugivorous New World primates that have rather hominoid-like teeth).

For the moment, though, let us assume that the similarities that do exist between pliopithecines and gibbons indicate an ancestral-descendant relationship or something close to it. This is one of the first steps to be taken in working out a phylogeny, or evolutionary sequence. Other factors to be borne in mind are relative dating (we wish to order the fossils from oldest to youngest) and geography (it would be unlikely that a South American Miocene primate was ancestral to any living Asian species). So, linking the fossils on the basis of their similarities, we can trace a hylobatid sequence that originates in Africa, probably during the Oligocene (Simons, 1965), and radiated, reaching Eurasia in the Miocene and becoming quite widely distributed there in Pliocene times. One (or more) of the Asian species was ancestral to living hylobatids. The geographical range of these became progressively restricted throughout the Pliocene and Pleistocene, probably as hylobatids became more strictly arboreal and as suitable habitats shrank.

Establishing such a hypothetical sequence is scarcely half the battle, for we are now faced with the problem of differences rather than similarities. If the phylogeny is correct, then certain changes have occurred, and it is necessary to describe and account for these changes.

Pliopithecines were gibbon-like above the neck but were very different in their postcranial anatomy. They had fore- and hindlimbs of more or less equal length, which implies that they were quadrupeds, not brachiators; this conclusion is supported by other postcranial features (Zapfe, 1958). Because pliopithecines were Old World primates, there has been a general tendency to compare them with the Old World quadrupeds, the cercopithecoid monkeys, and they have become known as "dental apes," creatures with hylobatid skulls and dentitions and cercopithecoid skeletons. Some workers have been reluctant to view pliopithecines as ancestral hylobatids because of the difficulty in accepting a transition from a cercopithecoid-like quadruped to a brachiator (Sarich, 1968). Accordingly, the dental and cranial similarities are regarded as parallelisms, that is, similar responses in different lineages to similar selective pressures.

The cercopithecoids are a relatively homogeneous group in locomotor terms, this being reflected in their postcranial anatomy, limb proportions, and so forth, and contrast markedly with the much more variable New World monkeys (Erikson, 1963). The evidence indicates that although the pliopithecines were indeed quadrupeds of sorts, they were much more like the arm-swinging quadrupeds such as the spider monkey or howler monkey. Thus, features of the hand, arm, forelimb joints, scapula, clavicle, and hindlimb all suggest that arm swinging comprised a significant part of the locomotor pattern, whereas leaping was somewhat less efficient than in quadrupedal leapers such as colobines.

The evidence does suggest a noncercopithecoid locomotor repertoire, and one much more like that of the spider monkey. If this is indeed the case, it makes it much easier to hypothesize a transition from arm-swinging quadrupedalism to brachiation as terminal-branch feeding by suspension became progressively more developed. The pliopithecines were evidently dimorphic in body size and canine size. If brachiation and lack of dimorphism are linked as has been suggested, then that change can be explained, too. Why did hylobatids become progressively more adapted to this specialized type of frugivorous feeding in the outer canopy? It has been suggested that competition in the Pliocene from the rapidly diversifying and more successful cercopithecines was a major factor in limiting the distribution and changing the niche of hylobatids (Napier and Napier, 1967).

THE GREAT APES

The other apes, the pongids, have often been described as "brachiators" or "modified brachiators" (Napier and Napier, 1967). However, the orangutan, although suspending itself from its arms, generally uses at least three extremities to support its weight; the chimpanzee, basically a quadruped, does arm swing but does not have an arm-propelled leaping phase in its locomotor pattern; the gorilla swings by its arms only when young, older animals being quadrupedal (Tuttle, 1967). Thus the term "brachiation," as applied to the gibbon, does not describe what pongids do any more than it defines spider monkey locomotion. Also, the term "modified brachiation" might imply to some that the pongids once passed through a stage in which their locomotion was gibbon-like. For these reasons, it is better to describe an animal's locomotion, if possible, with a label that is not too complicated (although this is difficult because locomotion, like all behavior, is a complex activity) and then list the anatomical correlates of behavior so that the behavior of fossil forms can be plausibly reconstructed.

The orangutan, *Pongo pygmaeus,* is confined today to a very restricted area of Southeast Asia, Borneo, and Sumatra and is a highly arboreal animal found in swampy rainforest (Davenport, 1967; Horr, personal communication). The population density of orangs has been reduced within recent years by human hunting, and there is evidence to indicate that they were much more widely distributed in Southeast Asia and China during the Pleistocene (Hooijer, 1948). Orangutans are frugivorous, feeding on fruits that are often located in the terminal branches. Because they are bulky creatures (male orangs may weigh more than 200 pounds), weight distribution is a problem. This is solved by using all limbs in climbing and feeding. Orang feet resemble the hands, having long curved phalanges and metatarsals and tiny first toes. Joints in the hindlimbs permit wide ranges of movements. The forelimbs are exceptionally long. Very similar stresses are placed on the orangutan shoulder girdle as on

Figure 16 Two orangutans, *Pongo,* in characteristic hanging posture. (Courtesy of D. Chivers.)

those of gibbons and spider monkeys, and consequently the morphology of trunk, thorax, and scapula is quite similar in these forms (Schultz, 1968).

The ancestry of the orang is poorly understood. Several pathways to the peculiar quadrumanous form of locomotion can be visualized: a gibbon-like or spider monkey-like ancestry might be possible, with perhaps the latter being somewhat more likely; recently, Walker (personal communication) has suggested the possibility of a knuckle-walking ancestry.

Little is known of the social behavior and organization of orangutans, though they apparently live in dispersed units consisting of mothers with infants to which adult males become attached only infrequently. Work with captive animals suggests that social behavior is (or was) more like that of chimpanzees than of gorillas (Reynolds, 1966).

The two African apes are confined mainly to regions of tropical rainforest in the Congo Basin, although some populations of chimpanzees live in open woodland in Tanzania. They have a similar type of specialized quadrupedal locomotion known as knuckle walking, in which the forelimbs contact the ground on the backs of the middle phalanges of the flexed fingers (Tuttle, 1967). There is a whole series of bony, muscular, and ligamentous adaptations to this posture. However, a number of features of the forelimbs, thorax, and trunk suggest a period in the ancestry of these apes when arm swinging and suspension formed a greater part of loco-motor behavior.

The gorilla (*Gorilla gorilla*) is restricted to tropical rainforest in lowland areas of the west Congo Basin and elevated regions in the east. Gorillas are ground-living foragers predominantly and con-centrate for the most part on tough herbivorous food (shoots, stems, bamboo) rather than on fruits. They are very large animals, females weighing up to 200 pounds or so, with the males twice as bulky. Large body size is probably important in defense against ground predators. Adult males rarely climb in the trees, although other gorillas do, and the large males sleep on the ground. Like the other pongids, gorillas build nests each night. Much of the day is spent eating and foraging for enough food to support their tremendous bulk. All the apes are nomadic, traveling daily to new feeding areas and nesting in a different site almost every night. The average home range of a gorilla troop is around 15 square miles.

The social organization and behavior of the eastern mountain gorilla have been studied by Dr. George Schaller (1963). Gorillas generally move in troops of between a dozen and twenty animals, each group containing a large dominant adult male (known as a "silverback" male because of the graying after a certain age of hair on the back) with a number of adult females and offspring. Other

Figure 17 Adult male lowland gorilla knuckle-walking. (San Diego Zoo Photo.)

THE ASCENT OF MAN

adult males may join the troop, but few appear to be as permanent as the silverback male, and they are subordinate to him. Single males are also frequently found wandering unattached to any group.

Gorillas are very dimorphic in body and canine size, probably because of the function of the male as protector of the more vulnerable females and also because of intermale competition. However, although silverback males are dominant to others (in the sense that they defeat other males in fights, determine when and where a troop moves, feeds, and rests, and serve as a focus of group attention), they are often tolerant of other adult males in the group—for instance, permitting them to copulate with females. Gorilla social units might then be regarded as single-male groups with other males loosely attached. This degree of tolerance is unknown in cercopithecoid species having one-male reproductive units and is perhaps possible because of the greater intelligence of such hominoids.

Gorillas are temperamentally phlegmatic creatures; the young tend to play alone a good deal, and adult time is taken up principally with eating rather than with personal intereactions. In these features they contrast strongly with chimpanzees. Stereotyped display behaviors, a lack of curiosity about strange objects, and the apparent absence of tool-using behavior (although this may be an artifact of too little observation in the wild) also separate gorillas from chimpanzees.

Chimpanzees are more widely distributed than gorillas and are found as well in forest-fringe and open-woodland habitats east of the Congo Basin rainforest (Reynolds, 1965). *Pan troglodytes* is found as a number of infraspecific variants from Sierra Leone in the west to Tanzania in the east. There is little sexual dimorphism in body size, males weighing on average a little more than 100 pounds and females somewhat less. However, canines are size dimorphic. South of the Congo River there is an enclave of the so-called pygmy chimpanzees that may constitute a separate species, *Pan paniscus.* These are of lighter build than *P. troglodytes;* little is known of their social behavior or detailed morphology. Although diminutive, these apes nevertheless still retain big male canines. Both species are knuckle walkers on the ground and in the trees, and they also are capable of both arm swinging and bipedalism.

The social organization and behavior of *Pan troglodytes* have been studied by several workers in a number of habitats including forest (Reynolds and Reynolds, 1965), woodland (Goodall, 1965), and savanna woodland (Itani and Suzuki, 1967). Chimpanzees are basically frugivorous animals, this being reflected in the relative proportions of the dentition. Fruit eating involves slicing through a tough outer cover to expose softer contents that require relatively little further chewing. Accordingly, chimpanzees have large, projecting incisors and relatively small cheek teeth. Most of the feeding takes place in the trees, so a considerable part of chimpanzee life is arboreal. They also nest in trees at night. However, when alarmed, chimpanzees will descend to the ground and escape by knuckle walking along terrestrial pathways. The diet is often sup-plemented with insects and small game, generally hunted and captured by the males. The catching of such animals, often smaller monkeys, is an occasion of great excitement and meat is a highly desired type of food, although eaten only infrequently.

Chimpanzees live in troops of around forty or fifty, which are fre-quently widely dispersed for feeding on ripe fruits. The troop splits into subgroups that consist of mothers and offspring, adult males and females, and all males. This pattern of organization resembles that of the spider monkey, also a ripe-fruit feeder, and is adapted

Figure 18 A chimpanzee intelligence test. When marbles are placed in the container, it descends, pulling up a food reward from inside the column. (Courtesy of Zoological Society of London.)

THE ASCENT OF MAN

Figure 19 Pygmy chimpanzee, *Pan paniscus,* and infant. (San Diego Zoo Photo.)

to a way of life requiring dispersion and aggregation. The troop may coalesce from time to time at particularly favorable feeding areas where fruit or nuts are abundant. Chimpanzees, unlike gorillas and orangs, are very noisy animals; quite often subgroups remain in contact by means of vocalizations, giving information about spacing, food, and predators. A group is frequently dispersed over several square miles. Home ranges of chimpanzees are larger than those of gorillas. Japanese primatologists studying the species in savanna woodland in Tanzania report group home ranges of up to 80 square miles or more.

Chimpanzees mature slowly, an infant remaining in close contact with its mother for at least 4 years. Kinship ties are strong, and female relatives in particular tend to stay together over long periods. It has been suggested that sexual relationships between mothers and sons do not occur and are avoided by the males' leaving the

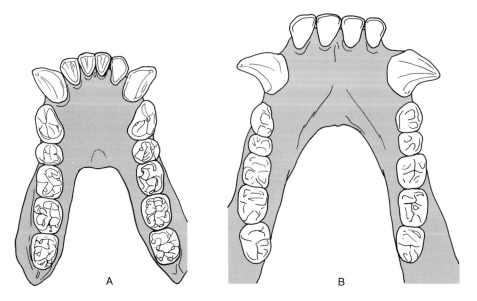

Figure 20 Primate dentitions reflect dietary preferences. The gorilla (A) and the chimpanzee (B) are drawn so that their premolar–molar series are of equal length. The herbivorous gorilla has relatively much smaller incisors than the frugivorous chimpanzee. (A: After Gregory, 1916. B: After Le Gros Clark, 1959.)

group into which they were born and joining another one. Adult males seem able to recognize their mothers even when they have been separated for some time; reunions are accompanied by highly excited greeting and display behavior.

Chimpanzees are emotional animals, much more extroverted than gorillas. They are curious about their environment, the young spending a great deal of time in exploratory and play behavior. They are also the most accomplished users of tools among the nonhuman primates and therefore provide a reasonable model for an examination of tool use among early hominids. Sticks and stones are used in displays and are thrown when an animal is standing or charging bipedally. Twigs are used for body scratching and leaves for wiping off mud. Tools are also utilized in food gathering, although their use is admittedly marginal, for chimpanzees are not dependent upon tools for survival in the way that hominids are. Stones may be used to crack nut shells, and chewed leaves have been observed in use as sponges to obtain water from hollow places in trees. At certain times of the year chimpanzees eat termites. They obtain them by fishing with small twigs or stems, scraping a hole in a termite mound, inserting the tool, and removing it covered with termites that are then eaten. Animals carefully select suitable materials for use in these activities, trim them to the correct length, and carry several with them when embarking on a termiting expedition. This behavior is learned by observation, young chimpanzees watching their mothers or siblings, and requires practice. It is a learned skill, not an instinct.

The chimpanzee's use of sticks in termiting is an excellent example of tool making, or object modfying. Unlike the making of hominid stone tools, the operations involved are very simple, standardization is not exact, and the tool is made for only one very specific purpose. Nevertheless, it does indicate how tools of this sort (as opposed to weapons) might have originated and suggests that the early hominids were tool users and modifiers at least to this extent.

Chimpanzee social behavior shows other resemblances to our own. Facial expressions, gestures, and postures are very similar. They indulge at times in display behavior analogous to dancing, the animals charging around together—frequently bipedally—vocalizing loudly. In zoos, chimpanzees have even been permitted to paint, and they do so with great pleasure, each animal developing its own style. They exhibit a rudimentary sense of composition and paint with as much skill as a human child does before reaching the representational stage.

This sort of evidence, together with that from comparative anatomy (Schultz, 1968) and biochemistry (Sarich, 1968) suggests that chimpanzees are the primates most closely related to man (along perhaps with the gorilla, whose behavior and morphology have become somewhat specialized), and therefore if we are to look for a model for early hominid social organization we should turn first to the chimpanzee. Of particular interest to us, apart from manlike behavioral characteristics such as tool use, facial expression, a latent esthetic sense, and so forth, is the flexible social organization characterized by both aggregation and dispersal, the existence of all-male groups that may wander widely in search of new food sources, the development of strong emotional ties between adults (the development of bonds of sorts between males being of particular interest), the fact that males may collectively hunt animal food, and the practice of occasional food sharing (primates are generally individualistic—selfish—foragers). These last two features are a function of the development of rudimentary cooperative behaviors of a sort that do not appear in other primates and that probably depended for their emergence in primates on the evolution of a sufficiently large brain (Reynolds, 1966).

A study of pongids and their ancestors can also yield clues about the evolution of other hominid characteristics, for example, the locomotor system. All three African apes are knuckle walkers. An analysis of their limb proportions is of interest in any reconstruction of their phylogeny. The intermembral index (a measure of the relative length of forelimbs to hindlimbs, expressed as a percentage) ranges around 117 in gorillas, indicating that arms are longer than legs. This (and other features) has led some primatologists to regard gorillas as "modified brachiators" even though they armswing only as subadults, and then only rarely. However, *Pan troglodytes* has a "lower" intermembral index (average 107), even though it is a smaller animal that arm swings more than the gorilla (Napier and Napier, 1967). The pygmy chimpanzee, a yet smaller species, apparently has a still lower index, barely over 100, indicating arms and legs of more or less equal length. As body size increases among these African apes, legs become shorter relative to trunk length (Walker, personal communication), and this causes the intermembral index to shift significantly. Clearly, the relatively long arms of the gorilla are not used in brachiating, nor need they

Figure 21 All apes have forelimbs which are long relative to trunk length, a feature presumably reflecting their common arm-swinging heritage. The gibbon (A) and orangutan (B) have remained arboreal and have the longest forelimbs, whereas the chimpanzee (C) and gorilla (D) have shorter arms, probably as adaptations to their more terrestrial way of life. All to same trunk length. (After Erikson, 1963.)

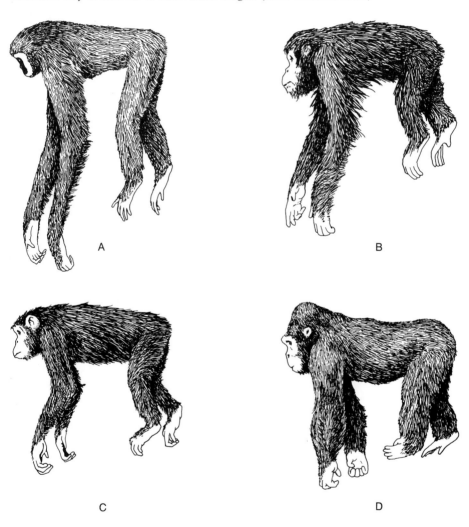

A

B

C

D

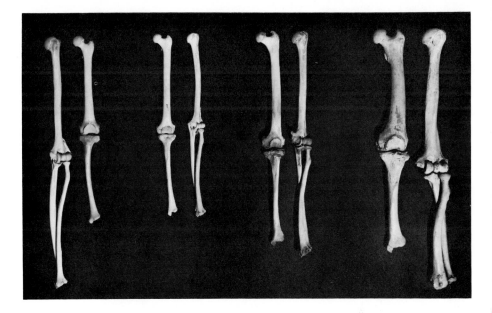

Figure 22 Forelimbs (humerus and radius) and hindlimbs (femur and tibia) of—from left to right—orangutan, pygmy chimpanzee, chimpanzee, and gorilla. Note that arms are relatively very long in the arboreal orangutan. Legs become relatively shorter in the knuckle-walking apes as body size increases. (Pilbeam and E. L. Simons.)

indicate a brachiating ancestry if it is assumed that a smaller gorilla ancestor looked more like a chimpanzee in body proportions.

Gorillas spend considerable periods of time sitting while feeding, stripping the vegetation around them before moving quadrupedally a short distance. They also feed with one hand while standing on the other three extremities. The relatively shorter legs of gorillas might be a result of this pattern of feeding, where some time is spent sitting. A similar feature (short legs) is found in gelada baboons that spend a great deal of their day feeding in a sitting position (Jolly, 1970). Also, a quadrupedal ground feeder needs to keep its center of gravity within the limits defined by the feet and one hand when the other hand is being used to collect food. Shortening the legs will shift the center of gravity slightly backward, thus stabilizing the body. Once again, a similar phenomenon is seen in habitually terrestrial cercopithecoids such as gelada baboons.

Therefore, we can conclude that the knuckle-walking ancestor of the gorilla and the other African pongids (if that ancestor was a knuckle walker) would have been a relatively small-bodied animal (around 50 to 70 pounds in weight) with arms and legs of approximately equal length. Knuckle walking probably developed when the ancestral African apes became relatively terrestrial (pos-

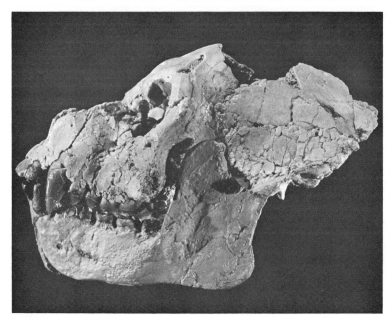

Figure 23 Side view of skull of *Aegyptopithecus zeuxis.* Cranium and mandible come from different individuals. The mandible is partially restored, as are the upper incisors. (Courtesy of E. L. Simons.)

sibly in response to competition from the omnivorous–frugivorous cercopithecines in the Miocene). The pre-knuckle-walking African apes were probably arm swingers, having arms relatively long compared with trunk length as well as the other adaptations of shoulder girdle, thorax, and trunk discussed above. Such an animal could have evolved equally well into the knuckle-walking apes, the long-armed orangutan, and bipedal hominids, for the locomotor repertoire of quadrupedal arm swingers such as spider monkeys contains a fair degree of bipedalism. This type of locomotion is preadaptive both behaviorally and anatomically to the development of habitual bipedalism. Alternatively, orangutan and human ancestors may have been knuckle-walkers.

Recently, Professor Charles Oxnard (1969a,b) has attempted to reconstruct mathematically the probable evolutionary pathways of the shoulder girdle in apes and man. He concludes that both man and the African apes passed through a stage in which the shoulder girdle was adapted to suspension and arm swinging, although it is likely that the common ancestor was too large to have "brachiated" in the sense that hylobatids do. This would imply that hominids did not pass through a knuckle-walking locomotor phase.

The earliest probable ancestors of the living hominoids occur in Oligocene deposits (around 30 million years old) in the Fayum region of Egypt (Simons, 1965). A few specimens were recovered during the first decade of this century, but many more fossil primates have since been discovered by expeditions led by Professor Elwyn Simons of Yale University. The oldest ancestral pongid whose exact provenance is known is *Aegyptopithecus zeuxis*, which has a probable age of around 28 million years. A number of jaws and isolated teeth are known of this species, and Simons' expeditions were fortunate enough to recover an almost complete skull, lacking the lower jaw and the upper incisors.

The skull is rather monkey-like, with a relatively small brain case and long projecting face, although the teeth are characteristically hominoid. A few postcranial bones are known and indicate that *Aegyptopithecus* was not a vertical clinger and leaper or a brachiator, but a quadruped, and one more like New World Monkeys than like cercopithecoids. The environmental setting of the Fayum region during the Oligocene was one of dense tropical rainforest lining the banks of sluggish, broad rivers. Possibly there were localized open areas, but these would not have been of any great extent. Presumably, *Aegyptopithecus* was an arboreal form.

If it is true that the earliest apes were frugivorous and the earliest African monkeys were herbivorous leaf eaters, then the Fayum hominoids might well have been utilizing a small-branch niche, quite possibly involving arm-swinging behaviors. Later hominoids, from the Miocene, have a postcranial anatomy that suggests arm swinging, and what is known of the living apes and monkeys suggests a considerable difference between the locomotor behavior of early hominoids and cercopithecoids.

Another Fayum species, *Propliopithecus haeckeli*, may come from

Figure 24 Reconstructed cast of *Propliopithecus haeckeli*. (Courtesy of E. L. Simons.)

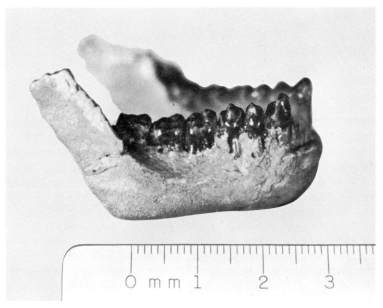

deposits older than *Aegyptopithecus* and might well be ancestral to the latter. It is known only from a couple of jaws and a few isolated teeth, and little can be said about it until it is analyzed in much greater detail.

DRYOPITHECINES

We can pick up the story of pongid evolution again a little later in time; fossil apes are known in large numbers from Miocene sites in Kenya and Uganda, the deposits ranging in age from around 23 million to some 18 million years (Simons and Pilbeam, 1965; Pilbeam, 1969a). A little is known about the paleoecology of these apes. Some sites were formed on the sides of active volcanoes, which were covered with dense tropical rainforest—a habitat not unlike that of the present-day mountain gorilla. Other collecting areas evidently sample a more lowland habitat and contain animals that lived in woodland and wooded savanna.

The Miocene apes are classified as a subgenus of the genus *Dryopithecus, D. (Proconsul)*. *Dryopithecus* was the name given to the first fossil pongid to be discovered, *D. fontani,* a Miocene form from France named in 1856. The first *D. (Proconsul)* species were found in the 1930s and described in detail by Sir Wilfrid Le Gros Clark and Dr. Louis Leakey in 1951. (When first described they were known by the generic name *Proconsul;* this was transferred in 1965 by Simons and me to *Dryopithecus* because we believed that species of the two genera were no more different than species within many living primate genera. This classification, it should be noted, is by no means universally accepted.) At least three species of *D. (Proconsul)* are known. Relative to the living pongids, they are primitive in teeth, skull, and postcranial skeleton. They are placed, with other species of *Dryopithecus*, in an extinct subfamily, Dryopithecinae; the living pongids are classified as Ponginae. The African Miocene dryopithecines could all have been derived from an earlier form such as *Aegyptopithecus*. However, by the Miocene the pongids were larger, with larger brains, and they had postcranial skeletons more similar to those of living pongids than did the Oligocene species.

At least one form, *D. (Proconsul) major,* may well be ancestral to a living ape, the gorilla (Pilbeam, 1969a). Remains of *D. major* come from sites that were on the forested slopes of volcanoes. The species was apparently quite sexually dimorphic, like the gorilla and unlike the chimpanzee. In body size it was probably about midway between the two present-day African apes. What is known of the face and dentition indicates a gorilla-like form, one less well adapted to the mastication of tough herbivorous food than that of the gorilla, although it possessed features suggesting tendencies in that direction. Parts of a vertebral column with resemblances to

CLASSIFICATION OF *DRYOPITHECUS*

Name	Earlier name	Age	Location
Dryopithecus fontani	*Dryopithecus fontani*	Middle Miocene– middle Pliocene	Europe
Dryopithecus sivalensis	*Sivapithecus sivalensis,*	Late Miocene– early Pliocene	Asia
Dryopithecus indicus	*Sivapithecus indicus*	Late Miocene– early Pliocene	Asia
Dryopithecus africanus	*Proconsul africanus*	Early Miocene	Africa
Dryopithecus nyanzae	*Proconsul nyanzae*	Early Miocene	Africa
Dryopithecus major	*Proconsul major*	Early Miocene	Africa

both gorilla and chimpanzee (and some to man, too) suggest a form with a reduced lumbar region (and therefore perhaps a shallow broad thorax) (Walker and Rose, 1968). The femoral remains of *D. major* indicate more agile locomotor behavior than in today's African apes (Le Gros Clark and Leakey, 1951), but a footbone (the talus) resembles quite closely that of the African forms and specifically is similar to that of the gorilla rather than the chimpanzee (Day and Wood, 1969). Additional undescribed material indicates the possibility that *D. major* was a knuckle walker (Walker, personal communication), although more evidence is required before this can be certainly established. *D. major* might also have been ancestral to the chimpanzee, but this seems somewhat unlikely. An alternative explanation for the gorilla-like features of *D. major* would be that they are simply parallelisms. I find this difficult to accept, but believe that further material is needed before we can be certain as to the correct phylogenetic relationships.

Another Miocene species, *D. africanus,* could have been ancestral to the chimpanzee. *D. africanus* was a small form, weighing perhaps around 30 or 40 pounds, and may well have been less restricted in habitat than *D. major,* because its remains are found in a variety of ecological settings (Pilbeam, 1969a). It was a quadruped, but evidently one capable of arm swinging, and may well have been a knuckle walker, too (Walker, personal communication; Conroy and Fleagle, *Nature,* in press).

It has been argued that these Miocene dryopithecines cannot have been ancestral to the living apes because they show no "brachiating" adaptations (Sarich, 1968). Implicit in this view is the assumption that the postcranial adaptations of the pongines are those of "brachiators." In fact, pongines do not brachiate in the way that gibbons do; rather, these adaptations are for knuckle walking, arm swinging, and arboreal climbing in which the arms are utilized to a greater extent than in quadrupedal monkeys. As I have noted, studies of the living apes point to a common pongid ancestor with arms and legs of subequal length, long relative to trunk length. Furthermore, the Miocene forms do show many features that resemble or foreshadow the living apes, and they do appear to have been quadrupeds capable of arm swinging, not cercopithecoid-type quadrupeds. In short, there are no good reasons why the Miocene species, or something like them, could not have been ancestral to present-day pongids.

Dr. Leakey (1967) has recently described one of the early Miocene African dryopithecines as a hominid ancestral to *Ramapithecus*, the late Miocene and early Pliocene form that is the earliest primate that can reasonably be regarded as a hominid. In fact, the early Miocene species, which we know mainly from incomplete jaws and isolated teeth, has characteristics typical of pongids and none that resemble hominids or might suggest it to have been ancestral to *Ramapithecus*. (Even if it were, classifying it as a hominid would still not be advisable.) Rather, the early Miocene form can be better viewed as close to the ancestry of the Eurasian dryopithecines (Pilbeam, 1968).

The earliest Eurasian *Dryopithecus* occur in deposits of middle Miocene age and later, ranging in age from 16 million to about 10 million years (Simons and Pilbeam, 1965). They are widely distributed, from Spain to China, and are classified into a number of rather closely related species. *D. fontani*, the first fossil great ape found, was a European form quite similar to *D. sivalensis*, known from India, Pakistan, and China. (*D. sivalensis* may well in fact include a good deal of another Indian "species," *D. indicus*.) The fossil history of the orangutan is obscure, though one of these Eurasian species might possibly have been related in some way.

Thus the fossil evidence points to a radiation of the apes in the Oligocene and Miocene, with chimpanzee and gorilla ancestors possibly being distinct as long ago as 15 million or 20 million years. Ancestral orangutans probably diverged even earlier. The pongid species ancestral to the hominids is as yet unknown or unrecognized, although probably when it is known it will be classifiable as a dryopithecine. As will be discussed in a later chapter, the hominids seem to have diverged from the ancestors of the other apes at least 10 million and probably 15 million years ago.

Recently, this fossil framework has been questioned by a number of workers, particularly those who have been involved with the study of the comparative biochemistry of primates. Drs. Sarich and Wilson (1968) have surveyed serum albumins (a blood protein) in a variety of living primates and have derived estimates of the relative "distance" between pairs of species. By making the assumptions that albumin has evolved at a constant rate in all lineages

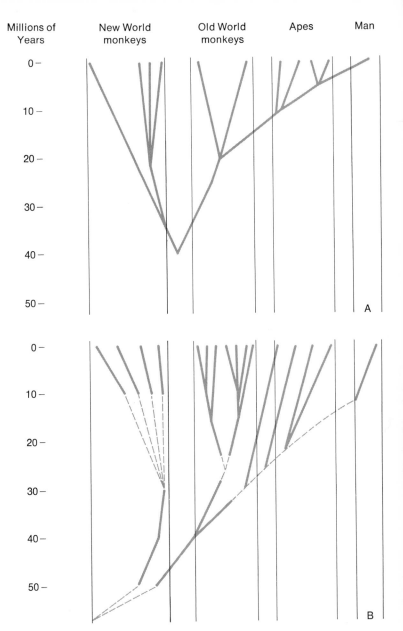

Figure 25 Comparison of the course of higher primate evolution as deduced from the biochemical analysis of albumin (A) and the fossil record (B). (Pilbeam.)

and that this rate can be measured, they have calculated the times of divergence of various lineages. Sarich and Wilson estimate that hominids, chimpanzees, and gorillas separated 4 million years ago, orangutans 7 million years ago, and gibbons 10 million years ago.

These "dates" do not agree at all well with estimates derived from the fossil evidence. There is reason to believe that some of the assumptions made in arriving at the biochemical ages may not be correct (Uzzell and Pilbeam, 1971) or that other factors might not have been taken fully into consideration. For the moment, let us assume that the fossil framework is acceptable.

What then can be concluded from a study of living primates about the protohominids of the Miocene? They were probably medium-sized arboreal apes (weighing 30 to 50 pounds), with fore- and hindlimbs of approximately equal length, relatively short stout trunks, broad shallow thoraxes, and shoulder girdles adapted to suspensory posturing and arm swinging. They may well have been knuckle walkers, at least incipiently so. They are likely to have been frugivorous forms and probably had a social organization not unlike that of the chimpanzee in which the social group was sub-divided for purposes of foraging into subgroups. The first hominids were probably creatures that came to the ground to feed at the forest fringe, in open woodland, and in grassland areas around lakes. The change of habitat and food type apparently resulted in profound changes in the masticatory apparatus and probably in the postcranial skeleton, too. Unfortunately, there are no open-country pongids to utilize as "models" for this phase of hominid evolution, but some of the large cercopithecoids can be so used in an attempt to explain the evolution of some of the basic hominid functional complexes.

Hominid Adaptations 3

In this chapter we shall discuss some of the unique adaptations of the hominids, especially those that can be detected in the fossil record. For the most part we shall concentrate on describing the character complexes that are found in the latest hominids but also shall briefly review some factors involved in their evolution, although these will be discussed more fully in later chapters. As outlined already, the three areas in which man differs most significantly from other primates are the dental apparatus (including jaws and face), the postcranial skeleton, and the brain.

THE DENTAL APPARATUS

The most noticeable dental difference between man and other primates is the absence in human males of large, projecting canine teeth (Le Gros Clark, 1964). Hominid canines are approximately equal in size in both sexes—although those of males are generally a little larger—and they do not project beyond the level of the other teeth. In both sexes, canines tend to be relatively smaller than those of even female nonhuman primates, and this is especially so for early hominids with their large cheek teeth. Perhaps as significant as the reduction in size is the fact that hominid canines have an altered shape. Instead of being sharp conical or bladelike teeth, they are chisel-like in form and resemble incisors. The incisors themselves are also small relative to the premolars and molars and have their crowns oriented vertically. Generally—at least in nonmodern populations of men—the cutting surfaces of upper and lower canines and incisors meet in an edge-to-edge bite. Incisors, canines, and premolars form a continuous series, there being no diastemata (gaps) between the teeth as there are in other primates (for example, between upper incisors and canines, or lower canines and first premolars).

Premolars and molars have rather rounded outlines, and their occlusal (chewing) surfaces have low, blunt cusps. These surfaces are covered with thick enamel—an adaptation to reduce wear in an animal that has an extended life span and (at least originally) eats tough food. In nonhuman primates the big upper canine shears against a single-cusped, rather bladelike front lower premolar; this is particularly true of males. In hominids the canines are small and

Figure 26 Upper dentitions of gorilla (*A*) and man (*B*) compared. In man, unlike the gorilla, incisors and canines are small and vertically oriented, canines are shaped like incisors, there are no gaps in the tooth row, the dental arcade is parabolic, and the cheek teeth are large relative to incisors and canines. (After Le Gros Clark, 1964.)

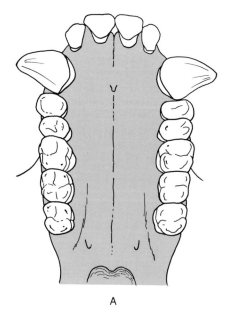

A

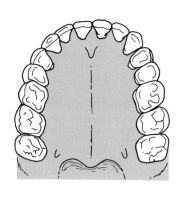

B

THE ASCENT OF MAN

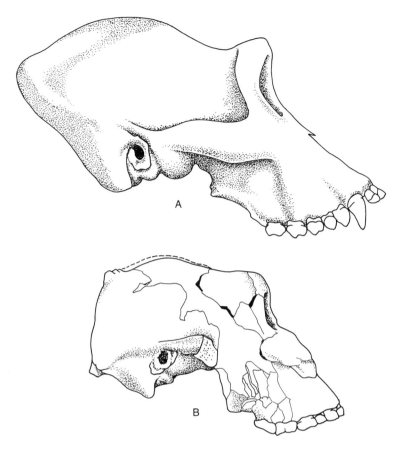

Figure 27 A robust ape, the gorilla (*A*), and a robust hominid, *Australopithecus boisei* (*B*). The hominid has small canines and incisors and a flatter face than the ape. Bony crests for the attachment of temporal and neck muscles are smaller in the hominid. (After Le Gros Clark, 1964.)

do not shear with the lower premolars. There has also been a tendency to "molarize" the premolars (that is, to add new cusps making premolars look more like molars), so the hominid first lower premolar has become a bicuspid tooth like the second premolar.

The teeth are set in thick, massive mandibles that are strongly buttressed on the internal side of the chin region, at least in earlier hominids (Scott, 1963). These features are adaptations to the stresses imposed by powerful rotatory and side-to-side chewing. The ascending ramus is high, because the teeth are carried in a short, deep face.

The outline of the tooth row, or dental arcade, is not ∩-shaped as it is in many primates. In some species of primates, especially males, the tooth rows on either side diverge toward the front. This is because of the presence of large canines and incisors. Because human front teeth are small, because the whole tooth row is fore-shortened (and also for other reasons), the hominid arcade is rounded at the front and the cheek-tooth rows are either parallel or divergent toward the back. The front part is gently curved so that the lower canines always stay in contact with the upper front teeth (canines or incisors) as the jaw is swung sideways during mastication. The backward divergence of the tooth rows may be a product of facial shortening, but it also brings each cheek tooth relatively a little closer to the jaw joint, thus slightly increasing the mechanical efficiency of each tooth during chewing.

As already mentioned, the face of hominids is short from front to back and deep from top to bottom, tucked in under the brain case. The ascending ramus of the mandible is high and more vertical compared with those of many other primates. These changes have occurred to improve the efficiency of mastication in hominids, particularly the chewing of tough vegetable food, for some of these changes are analogous to those that have occurred in other lineages of herbivorous mammals (Crompton and Hiiemäe, 1969). Shortening the dentition as a whole and shifting the dentition backward relative to the condyles (reflected in the vertical ascending ramus of the lower jaw) help to increase the mechanical efficiency of the dental apparatus by reducing the load arm around the jaw joint at all parts of the tooth row. This is further aided by shifting the origin of the masseter muscle forward (by shortening the face) and lowering its insertion (by deepening the face), thus increasing its moment arm around the jaw joint. The main part of the temporal muscle is best developed anteriorly and pulls relatively vertically (and is attached to a vertical anterior border of the ascending ramus). Its moment arm is also increased. The result is very powerful mastication, an activity that involves a good deal of lateral movement (due to the antagonistic activities of the masseters and pterygoids) as well as fore and aft shifting. Therefore, chewing in hominids involves powerful movements of the lower teeth on the uppers in just about any direction or combination of directions, activities facilitated by the absence of projecting and interlocking canines.

Thus the morphology of the dentition and face has been profoundly affected by selection pressures acting to produce powerful chewing. The basic hominid dental adaptations seem to have been established very early in hominid phylogeny; possibly they were "the" basic adaptations. Factors affecting their evolution are discussed below. The main trends in hominid dental evolution during the Pleistocene have been toward reduction in size of the cheek teeth, with little or no change in size of the incisors and canines

Figure 28 (opposite) Muscles of mastication: The temporals, masseters, and medial pterygoids (A and B) are primarily jaw-closing muscles, whereas the lateral pterygoids (C) move the jaw from side to side as well as pull it forward. (After Campbell, 1966.)

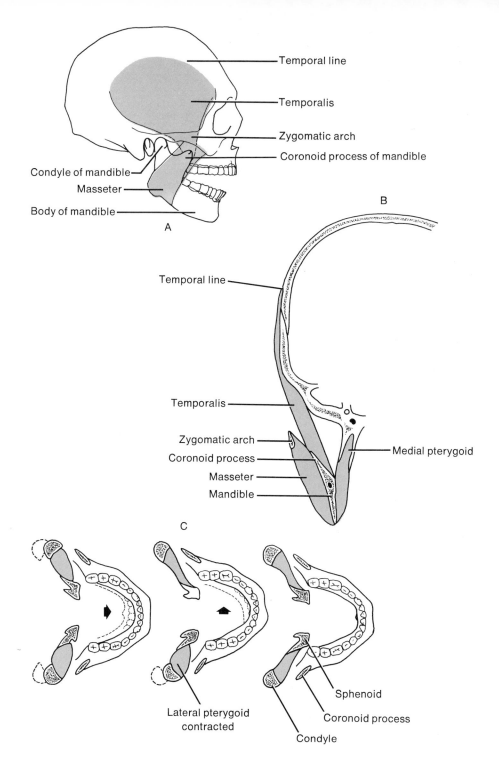

A

Temporal line

Temporalis

Zygomatic arch

Coronoid process of mandible

Condyle of mandible

Masseter

Body of mandible

B

Temporal line

Temporalis

Zygomatic arch

Coronoid process

Masseter

Mandible

Medial pterygoid

C

Lateral pterygoid
contracted

Sphenoid

Coronoid process

Condyle

Figure 29 A male mandrill, *Mandrillus sphinx*. (San Diego Zoo Photo.)

(Le Gros Clark, 1964). This may well be due to the introduction of greater and greater amounts of meat into the diet and also to the use of fire to cook food, thus making it less tough.

Recently, Dr. Clifford Jolly (1970a, 1970b) has drawn attention to some interesting parallels between gelada baboons and hominids in dental and cranial morphology and has suggested that these parallelisms might be due to adaptations to broadly similar diets. In this chapter we shall discuss Jolly's ideas by way of introducing an approach: that structure is best understood in terms of function, and function is best understood within the broader setting of behavior and ecology. Such an approach to living primates can aid enormously in making sense out of the hominid fossil record. The living primates to be considered here are the large African cercopithecoids of the genera *Mandrillus* (mandrills and drills), *Papio* (ordinary baboons), and *Theropithecus* (gelada baboons).

Mandrills and drills live in the tropical forests of West Africa. They are mostly forest-floor animals, although they can climb trees well, and they feed on fruits, tubers, and insects. Baboons of the genus *Papio* (which comprises at least two species, and perhaps more, but we shall concentrate on *P. cynocephalus*) are widely distributed in a variety of African habitats ranging from tropical rainforest to semidesertic scrubland. They are eclectic feeders, eating all types of vegetable food from fruits to grass, depending upon habitat. They also eat meat. Baboons have been regarded as open-country animals, but there is reason to believe that they are basically forest-fringe and woodland creatures. A number of workers believe that the open savannas of present-day Africa are a relatively recent phenomenon. Jolly believes that open wooded areas were until recently much more widespread but that trees have been lost because of the agricultural activities of man. The only truly open grassland areas, he believes, would have been around lakes that flooded seasonally, thus inhibiting tree growth. Therefore, *Papio* baboons should be thought of as unusually versatile creatures that have been able to move, because of their adaptability, into grassland niches as these opened up. Fossil baboons are found in many late Pliocene and early Pleistocene sites but occur with lowest frequency in deposits for which a lakeside (that is, open-country) environment can be inferred.

Gelada baboons are confined today to the high plateau region of Ethiopia, sleeping on steep rocky cliffs and feeding in open grassland. Their diet consists of grass blades, seeds, rhizomes, small herbaceous bulbs, leaves, and arthropods. This food is scattered evenly over the ground (unlike tree products), and geladas feed in a sitting position, picking up these small objects with their hands (which have short index fingers capable of being used with the thumb in a fine-precision grip) before transferring a fistful to the mouth. The morsels are relatively unnutritious, and feeding occu-

Figure 30 Male and female gelada baboons, *Theropithecus gelada*. Note the short face. (San Diego Zoo Photo.)

pies most of the day. Geladas shift position either by shuffling along on their bottoms or by waddling short distances bipedally. Fossil geladas are known in large numbers from Pliocene and Pleistocene sites in East and South Africa and are particularly abundant in lakeside settings. Some were enormous animals, for cercopithecoids. Most of them became extinct in the late Pleistocene, probably because of the ease with which such large, slow-moving primates could be hunted by man. The Ethiopian geladas are therefore a last, relict population.

Let us next briefly examine the morphology of the dentition and skull in each of these genera, before trying to explain the reasons for the morphological contrasts. Briefly, the dentition can be sub-divided into three regions: the incisors, which prepare food; the cheek teeth, which process food; and the canines and front lower premolars, which are primarily involved in display behavior and fighting. In *Mandrillus* the incisors are broad and high-crowned and the cheek teeth are relatively small with rounded, low occlusal surfaces. Incisors become worn down rapidly, more so than molars and premolars. In the large males the incisors are increased in size, but cheek teeth are barely longer than in females.

Theropithecus dentitions provide a very different pattern, with large premolars and molars and small incisors that are little worn during life. The molars have deep infoldings of enamel, high pointed cusps, and large ridges at the front and behind. As the teeth wear, they expose alternating bands of dentine and tough enamel, thus pro-viding an efficient mincing surface. In the much larger extinct geladas of the Pliocene and Pleistocene, the cheek teeth were pro-portionately larger, but the incisors were no larger than in living species. *Papio* is intermediate in dental proportions between the other two genera. *Mandrillus* has proportionately the largest canines and *Theropithecus* has the smallest, a phenomenon that Jolly believes may be either linked genetically with the small incisors or a conse-quence of muzzle shortening (in turn associated with a number of factors including incisal reduction).

Cranial shape also varies in these three genera. In *Theropithecus* the face is flat and deep, the ascending ramus of the mandible high and vertical. *Mandrillus* has a long, projecting face, and the mandib-ular ramus is set at a wide angle to the body of the mandible so that the jaw joint is barely above the plane of the cheek teeth. *Papio* is intermediate. The sagittal crest for the temporal muscle is set well forward in *Theropithecus*; the zygomatic arches are strong and circumscribe a large temporal fossa, which thus produces a marked postorbital constriction. Such features are less well devel-

Figure 31 (opposite) Comparison of skulls and lower dentitions of female *Papio* (A and C) and female *Theropithecus* (B and D). Note the flatter and deeper face of the gelada and the higher and more vertical ascending ramus. Considering the dentitions, the gelada has much smaller incisors and canines compared with cheek teeth than does the ordinary baboon. (After L. S. B. Leakey and T. Whitworth.)

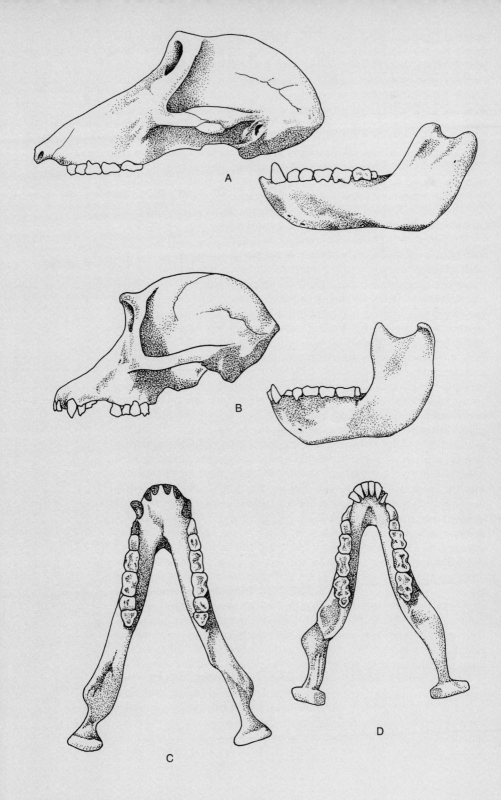

oped in *Papio* and still less in *Mandrillus,* where, for example, the crests for the temporals are set well back on the brain case. The temporal muscles in *Theropithecus* are powerfully developed anteriorly, hence the large temporal fossas and anteriorly developed crests. Their pull is much more nearly parallel to that of the masseter muscles and pterygoids than in the other monkeys. The strong zygomatic arch supports a powerful masseter muscle. The increased height of the ascending ramus of the mandible provides a greater area of attachment for the masseters and pterygoids. Greater height also increases the moment arm of the masseter around the jaw joint, thus raising bite power. *Mandrillus* has none of these features. Of particular note, in *Mandrillus* the temporal muscles pull at an oblique angle to the tooth row, probably serving to move the incisors back and forth against one another during nibbling.

The dental and cranial features are adaptations to differences in diet in the various species. The food eaten by *Mandrillus*—fruits and tubers mainly—requires preparation with the incisors to remove the tough outer coat, whereas the pulpy interior needs little mastication. Geladas, on the other hand, eat large amounts of small, tough morsels that do not have to be prepared incisally but rather crushed and chewed by the large cheek teeth. The large masseters and pterygoids can exert considerable force on the food; because they effectively carry the deep mandible in a sling of powerful muscle, side-to-side shearing movements can be produced. This is what all mammalian herbivores can do, and the gelada is simply a very good herbivore. However, it is more efficient than the average herbivore because the temporal muscle, best developed anteriorly, pulls almost vertically. Compared with the *Mandrillus* temporal, the moment arm of the *Theropithecus* temporal is greater, another adaptation for more power. However, the greater power is applied in such a way (because of the verticality of the average vector of the muscle) that this power is "controlled" power, most effectively applied in moving teeth from side to side. Thus, gelada mastication is highly efficient for chewing tiny, tough vegetable items. Because of the heavy use of the cheek teeth, they become packed close together during life. This packing occurs because the teeth "drift" forward during eruption as pressure is applied to them. This results in a more continuous cutting surface, and moreover one in which the gums are protected because food is less likely to become lodged in spaces between the teeth. The first molar becomes rapidly worn before the second erupts, and this is well worn in turn before the third molar comes into occlusion—by which time the crown of the first may be almost completely removed. This extreme "differential" wear is uncommon in other monkeys.

Geladas exhibit adaptations to sitting while feeding. The hindlimbs are relatively short, and the fat-filled cushions on the buttocks are also related to this posture. Most Old World monkey females have sexual skin around the genitalia that changes color when the female is in estrus and sexually receptive. In geladas the female sexual skin is on the chest, and Jolly believes that this is again associated with the fact that feeding, taking up so much of the geladas' day, is done in a sitting position.

The digits of hands and feet are short in *Theropithecus*, an adaptation to ground living that reduces shearing stresses on the phalanges during quadrupedal locomotion. The index finger is particularly short in geladas, giving a high "opposability index" (ratio of thumb to index finger length), and this is undoubtedly related to the requirements of picking up small objects.

Hominids also have dentitions with small incisors and large cheek teeth that become packed together and show strong differential wear. Their faces too are flat and deep; the mechanics of muscle action are broadly as in geladas, although the small canines allow more rotatory movements during tooth occlusion. Jolly has suggested that these morphologies occur because the basic hominid dental adaptation is to chewing large amounts of small, tough items, the type of food that would be collected mainly on the ground. As a generalization, this is a highly significant and important suggestion. The differences between geladas and other large cercopithecoids are analogous to those differentiating hominids from pongids and presumably developed as correlates of a shift from arboreal, mainly frugivorous feeding in forests to terrestrial foraging at the forest fringe, in woodland, and in open grassland.

However, we should not overlook the differences between geladas and hominids. Hominid incisors are sharp, bladelike cutting teeth, with crowns oriented vertically (Every, 1970). Canines resemble incisors morphologically and form with them a continuous anterior cutting blade. This is set in a short face, which means that the mechanical advantage of these teeth is relatively greater than in long-faced forms, an adaptation for powerful slicing. The dental arcade at the front of the face is gently rounded so that when the lower jaw moves from side to side the edges of the lower canines retain contact with the upper incisors; slicing can occur in both up-and-down and side-to-side movements.

Hominid molars have thick enamel, covering low rounded cusps that soon become truncated by wear, leaving the crowns flat and featureless even before the enamel is perforated (Butler and Mills, 1959). The outlines of the cheek-tooth crowns are rounded, so that at all times during rotatory occlusion a tooth edge is shearing across a flat surface, slicing the food before it is ground between the occlusal surfaces.

Further adaptations for powerful chewing include the shifting back of the tooth row so that it comes to lie partly under the frontal region of the brain case. This reduces the load arm on any particular tooth in the lower jaw (when considered as rotating around the jaw joint), and this serves to increase the mechanical advantage of all teeth.

Thus early hominids may well have been chewing small items, but they were preparing food with incisors and canines in a unique way and processing it in a similarly unusual manner with the premolars and molars. What the diet of the earliest hominids was

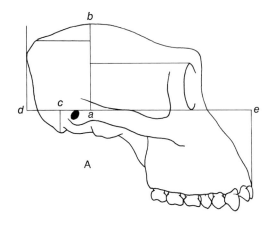

A

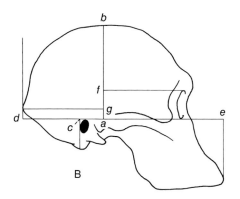

B

Figure 32 Skulls of gorilla (*A*) and *Australopithecus africanus* (*B*) showing major differences in cranial forms. The index *cd/ce* reflects the position of the foramen magnum, *ag/ab* the extent of the neck-muscle mass, and *fb/ab* the height of the brain case above the eye socket. The graphs (*C*) show values for population means and ranges of these indexes. (*A* and *B*: After Le Gros Clark, 1964. *C*: After Tobias, 1967.)

C

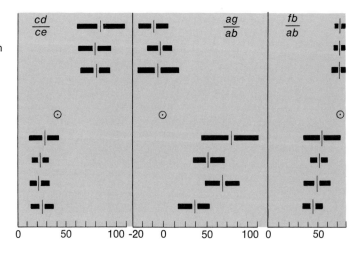

comprised of can only be guessed at—probably seeds, roots, and other such items, possibly tough fibrous shoots, and perhaps also meat and even bone.

This brief account of cranial and dental mechanics in primates has shown, hopefully, that it is possible to interpret fossil specimens—often quite fragmentary ones—in a functional way and that this enables primatologists to say much more about the behavior of extinct species. It is behavior, after all, that selection acts upon to produce evolutionary change.

The other part of the hominid skull—the brain case—differs considerably from that of other primates, and these differences increase as the hominids are traced through time. The morphology of the brain case is affected mainly by the size and shape of the brain. Paleontologists have described a number of features that distinguish the skulls of humans from those of nonhominid primates (Le Gros Clark, 1964). There have been three principal characteristics on which attention has been concentrated, and these have been defined by indexes. The first index reflects the relative position of the occipital condyles, the parts of the skull that articulate with the vertebral column. In modern man the condyles are relatively far forward, whereas in apes they are much further back. Earlier hominids are intermediate. It is not certain exactly what this index indicates: possibly head balance, possibly facial reduction in hominids, possibly posterior and inferior brain expansion, possibly a combination of these and other unknown factors. A second index is the relative height of the neck muscles on the occipital bone at the back of the skull. The muscles are smaller in hominids, including late Pliocene ones, and do not reach as far up the skull as in apes. This is probably related to the facts that hominids have nonprojecting faces and that the males do not have large canine teeth for displaying and fighting with, which would require powerful neck muscles. Finally, an index has been devised that reflects the height to which the brain case rises above the top of the facial skeleton, as defined by the upper margin of the eye socket. In man this index is high, in apes low; earlier hominids are intermediate or fall within the ape range. Brain expansion would affect the index and so too would the way in which the brain case is hafted onto the facial skeleton, their relative positions probably being determined by mechanical considerations. Once again, this index is difficult to evaluate, for either factor might be involved, not necessarily both together, and two skulls might give the same index for totally different reasons. This only points up the necessity for great care in the use of indexes; it is desirable to be absolutely sure about the significance of what is being measured.

Man shares a number of postcranial features with the other hominoids that probably reflect his evolution from an armswinging ancestor (Schultz, 1968). These include shallow broad thorax, reduced lumbar region, long clavicles, scapulae on the back of the thorax, and long arms relative to trunk length. Certain features of the shoulder, elbow, and wrist joints also attest to this ancestry. However, at least some of these features have been retained because hominids have become erect bipeds (for example, thoracic shape, scapular disposition, long clavicles), and there have been numerous morphological changes that have accumulated since hominids became bipedal.

The main changes in upper limb morphology are due to the fact that the human hand and forelimb are no longer utilized in locomotion but instead have become involved almost exclusively in manipulating objects—tools, weapons, and other parts of the environment (Oxnard, 1969a,b). Although human arms are long relative to the trunk (a ratio of about 110:100), they are shorter than those of gorillas and chimpanzees (both with a ratio of approximately 120). It should be borne in mind that the human lumbar region is longer than that in any of the living pongids, and this would affect the ratios. Forelimbs of early hominids were probably intermediate in relative length between modern man and the extant apes. The scapula and clavicle of man differ from those of all other primates. They are adapted to forelimb mobility and are not for use in either quadrupedal locomotion or arm swinging. However, according to theoretical work recently completed by Dr. Charles Oxnard (1969a,b), the human scapula and clavicle could be derived most easily from those of the orangutan, a large-bodied, arm-

Figure 33 (opposite)   Adaptations to bipedalism. A: Vertebral columns of an ape (1) and man (2). In man there is a distinct lumbar curve, and that part of the vertebral column is set at a sharp angle to the sacrum. B: Femurs of an ape (1) and man (2). In apes the long axis of the femoral shaft is parallel to the vertical axis, whereas in man there is a "carrying angle." In man weight is borne mainly on the outer condyle, which is larger than the inner condyle. In apes weight is borne more on the inner condyle. C: The big toes of apes (1, 2, and 3) are divergent and capable of grasping. In man (4) all toes are short, and the big toe is set parallel to the others. Weight is transmitted, as shown by the dotted lines, through the big toe, which is very important in bipedal walking. D: Contrasts between the form of the metatarsals in a gorilla (1) and man (2). The bases of the metacarpals are indicated in transverse section by broken lines, the heads by continuous lines. In man the heads contact the support, whereas the bases form a raised transverse arch, a spring under tension, which is of great importance in bipedal walking. (A and B: After Campbell, 1966. C and D: After Le Gros Clark, 1959.)

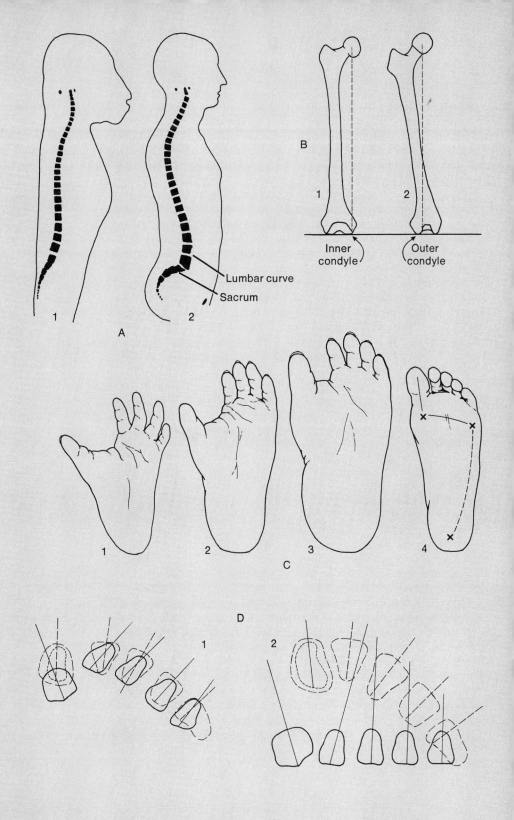

Lumbar curve

Sacrum

A

Inner condyle

Outer condyle

B

C

D

swinging form. Interestingly, the fragments of shoulder girdles known from early hominids resemble in certain features those of the orangutan. Apparently, the shoulder girdle in man has changed because of the loss of arm-raising and suspensory behaviors.

The human hand has become extremely well adapted to object manipulation. Fingers are relatively short, and the thumb is relatively long and capable of being rotated so that the pulps of thumb and fingers can be approximated. Early hominids had shorter thumbs and more curved finger bones than modern man, probably because their hands were adapted for somewhat less fine manipulative ability (Tuttle, 1967).

Those parts of the brain that control hand movements are far better developed in man than in apes. For example, the area of the motor cortex devoted to the hand (particularly the thumb and index finger) is greatly expanded, as are the regions in the cerebellum (the part of the brain responsible not so much for the movements themselves as for the control and integration of the movements). It has been argued that the tool-making and general manipulative abilities of hominid hands can be deduced from a study of hand proportions, joint morphology, and muscle organization (Napier, 1962a). However, although human hands have clearly evolved to be more adept, it seems more likely that in this case it is the central (brain) factors that are more important than the peripheral ones.

The morphology of the remainder of the human skeleton is affected mostly by the demands of habitual upright bipedalism. Human walking is a complex process unique among mammals in that it involves only the hindlimbs, which contact the support alternately, with the trunk held erect and fully extended on the hindlimb (Zihlman, 1967). During certain phases of walking, both hip and knee joints are fully extended. Weight is transmitted down the leg and along the outside of the foot to the big toe, which is generally the last part to contact the support during the propulsive phase of walking or running. During walking, the body's center of gravity shifts very little in any plane, and this restriction of movement means that a minimum of energy is expended. Clearly, bipedal walking requires complex central nervous system control. However, peripheral adaptations are much more obvious than in the case of the forelimb.

Walking involves a whole series of flexions and extensions of various joints, along with pelvic rotation and tilting. Although we need not go into complete detail here, as examples we can cite the following: During walking, as one leg is lifted off the ground, the

Figure 34 (opposite) Functional maps of the motor cortex in a monkey (*A*) and man (*B*). Compared with the monkeys, certain cortical areas in man have expanded differentially, particularly those associated with the hand and face. Presumably, these changes reflect the evolution of tool making and language in hominids. (*A*: After Washburn, 1960. *B*: After Wilder Penfield and his associates at the Montreal Neurological Institute.)

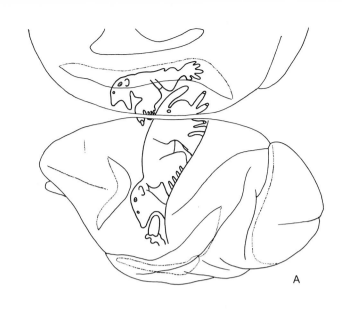

A

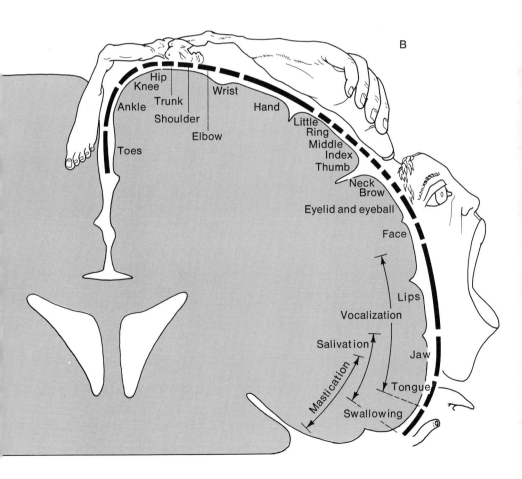

B

65

pelvis rotates around the leg that remains in contact, thus moving the free limb forward. As the leg is raised, the pelvis tends to collapse but is prevented from doing so by muscular contraction.

Let us now outline briefly some of the adaptations of muscles, bones, and joints to bipedal walking. The small gluteal muscles—gluteus medius and gluteus minimus—run from the side of the iliac bone of the pelvis to the top of the femur. Their contraction prevents the pelvis from slumping to the unsupported side when only one foot is in contact and also rotates the pelvis around the fixed limb during walking. This is possible because of the relative dispositions of the muscle origins and insertions. The gluteus maximus (the large muscle of the buttocks) acts as a hip extensor, originating on the back of the pelvis and sacrum and running to the back of the femur. It also serves to check forward momentum because the advancing limb acts as a brake on coming into contact with the support. In living apes these muscles act mainly as abductors; that is, they swing the leg outward. In bipedal walking the gluteus maximus cannot act as a hip extensor, nor do the small glutei counteract pelvic tilting or provide rotation.

There are also numerous bony and ligamentous adaptations to walking, a few of which will be mentioned here. The human vertebral column has developed a series of curves, the most marked of which is the lumbar curve, which allows the hindlimb to be in a fully extended position while a vertical trunk position is retained. The pongids have no lumbar curve, and the region is shorter and more inflexible than in man. The human lower limb has elongated greatly compared with that in apes, an adaptation to increasing stride length. The iliac part of the pelvis is broadened, and the upper surface forms a sigmoid curve that brings the origin of parts of the small glutei in front of the hip joint, thus ensuring that they will act as hip rotators during walking. Iliac blades and ischium are aligned at a more acute angle than in apes because of tilting of the ilium associated with the acquisition of truncal erectness. The human ischium is short, thus reducing the moment arm of the hamstrings, muscles that extend the hip and flex the knee. This reduction is probably associated with the development of the powerful gluteus maximus as the main hip extensor.

In man the center of gravity passes a little behind the hip joint and through or just in front of the knee, thus ensuring that both joints will be kept in an extended position during standing with a minimum energy expenditure. Hyperextension at the hip joint is prevented by the development of a strong iliofemoral ligament

Figure 35 (opposite)   In man (A) the center of gravity lies just behind the midpoint of the hip joint and in front of the knee joint so that both hip and knee are extended in standing, thus conserving energy. In the chimpanzee (B) the center of gravity falls within the rectangle formed by the four extremities. When the ape walks bipedally (C), its center of gravity is displaced from side to side and up and down. In man the center of gravity is displaced only a small amount; this makes human walking very efficient. (A–C: After Zihlman, 1967.)

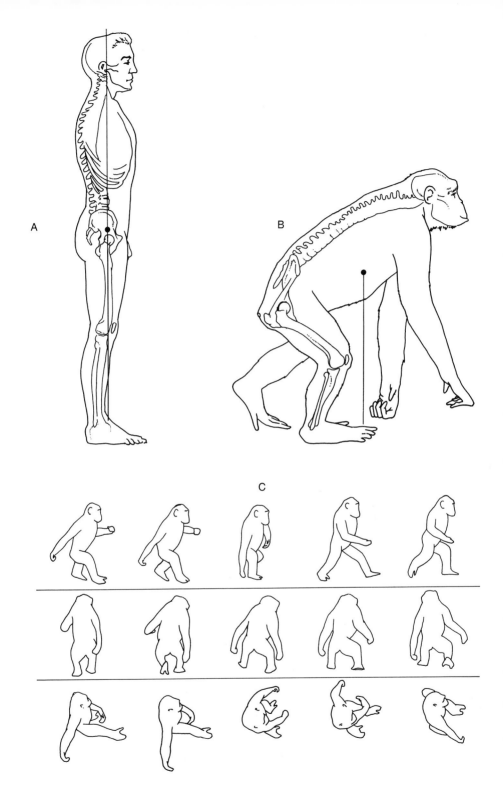

A

B

C

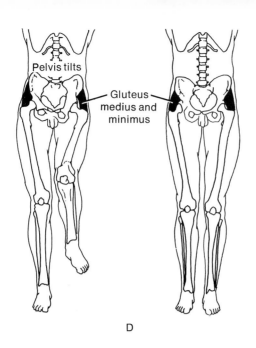

Pelvis tilts

Gluteus
medius and
minimus

D

E

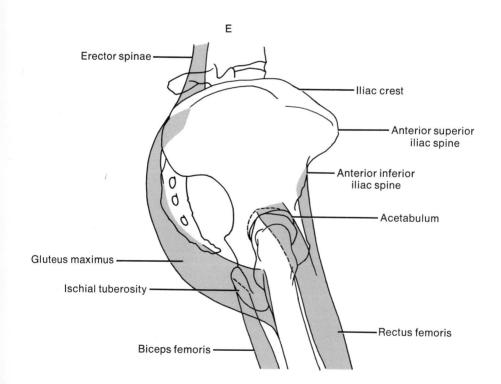

Erector spinae

Iliac crest

Anterior superior
iliac spine

Anterior inferior
iliac spine

Acetabulum

Gluteus maximus

Ischial tuberosity

Rectus femoris

Biceps femoris

Figure 35 (continued, opposite)   Important features and movements of the human hip joint are shown in *D* and *E*. The small gluteal muscles (medius and minimus) play a vital role in balancing the trunk on one leg when the other is raised off the ground. This they do by contracting, thereby preventing the pelvis from collapsing on the unsupported side (*D*). These muscles also rotate the pelvis around the fixed limb, bringing the free limb forward for the next stride. In man one of the most important pelvic muscles is the gluteus maximus (*E*). This is a powerful extensor of the hip and is involved in walking and in standing up from a sitting position. In apes the gluteus maximus is only poorly developed and pulls mainly across rather than behind the hip joint. Thus it is not a major hip extensor. (*D*: After Doug Cramer's drawings in Zihlman, 1967. *E*: After Campbell, 1966.)

that leaves prominent markings on both bones. This feature can be observed in early hominids.

The acetabulum and the femoral head are large in man, reflecting their importance in weight transmission. The acetabula face outward and forward. In standing, the knees are close together, and because the hip joints are further apart the femur and tibia do not form a straight line as in apes. In this way, during walking, the foot, tibia, and knee joint of each leg stay close to the line followed by the center of gravity, and thus there is less energy expended in maintaining the center of gravity above the supporting limb. Because of the femoral angle, essentially the upper part of the lower limb and the pelvis rotate around the tibia and foot during forward progression (although some rotation occurs at hip, knee, and ankle joints too).

The knee joint is adapted to bear weight through the lateral condyles, which are larger than the medial condyles. This is because in man the knees are close together and the load of the body's weight passes through the outer side of the knee joint, rather than inside it as in the case of apes, where the legs are carried far apart. The main ligaments of the knee joint—the cruciate ligaments—are so arranged that they are under tension no matter what the degree of flexion or extension of the joint. This is achieved by the presence of a deep notch between the femoral condyles and the ability of the condyles to slide back and forth on top of the tibia in various stages of movement of that joint.

The tibial shaft is straight in man, not curved laterally as in apes, and the plane of the joint between tibia and talus is horizontal. Movement at the tibiotalar joint is mostly in one plane—flexion and extension. The joints of the foot permit much less internal mobility than in apes, an adaptation for ground walking rather than for climbing on supports that may be oriented at various angles.

The human foot has a longitudinal arch like other primates do but is unique in the development of a transverse arch, formed because of the shape of the bony elements and supported by ligaments and tendons (Napier, 1967). These arches convert the foot

into a complex spring under tension and allow the foot to transmit the stresses involved in walking both when body momentum is checked as the foot first strikes the ground and when the foot is used in propelling the body. The metatarsals are relatively short and straight, the first and fifth being the most robust, reflecting the manner in which weight is transmitted along the outside and across the ball of the foot, principally to the big toe. All toes are short, and the big toe is particularly robust. Unlike those of the apes, human feet are not involved in grasping; the first metatarsal and toe lie parallel and close to the others. Associated with this change are alterations in joint morphology, ligaments, and so forth.

Outlined here is only a partial list of human adaptations to bipedal walking. Although late Pliocene and early Pleistocene hominids were not as well adapted as middle and later Pleistocene forms to bipedalism (Zihlman, 1967), they were nevertheless clearly bipeds and not quadrupeds or armswingers. Exactly when bipedal locomotion became established in hominid populations is unknown: probably in the middle or early Pliocene, possibly after the dental adaptations had occurred. As noted already, the locomotor repertoire of apes includes some bipedalism—much more than in monkeys—although their upright walking is far less efficient than man's.

What selection pressures might have produced the changes? Apes are bipedal in a variety of circumstances: When scanning surrounding areas, while carrying food or other objects, or when engaging in various types of displays. Probably hominids became ever more bipedal as they were required to carry tools and weapons while hunting and foraging, to use weapons in killing game, and to carry food, infants, and other possessions from one place to another. The final evolution of characteristically human bipedalism was probably bound up with increasing home-range size as hominids ultimately became highly effective and mobile big-game hunters.

THE BRAIN AND BEHAVIOR

Perhaps the most profound differences between man and other animals are behavioral. Complexity and organization of behavior are directly related to the fact that the brain of *H. sapiens* is greatly enlarged compared with the brains of other primates. In considering the evolution of the hominid brain, we have to rely on indirect evidence (Holloway, 1968), for there are unfortunately no brains fossilized so that their internal structure is preserved. Available are external casts of brains or internal molds of brain cases, neither of which are particularly satisfactory, for they can provide information only about the grosser aspects of external brain morphology.

Another source of information is study of the internal structure and organization of brains in living primates. Thus, ape brains can be used as "models" for early stages in hominid brain evolution, although this can be done only with extreme care, for the living nonhuman primates are themselves products of long periods of independent evolution. For example, the chimpanzee brain may not resemble particularly closely the brain of the common ancestor of the African apes and man. The final type of behavioral data available is the information that can be gleaned from studying teeth, jaws, postcranial bones, stone tools, associated animal remains, settlement patterns, burials, and cave paintings. In a sense this is all fossilized human behavior.

The human brain is approximately three times as large as that of the largest nonhuman primate. In modern man the average brain volume is around 1,400 cm^3, whereas that of the gorilla is a little over 500 cm^3; the chimpanzee mean volume is around 400 cm^3. In particular, the cerebral hemispheres are greatly expanded in man, but it must not be forgotten that other important parts of the brain have expanded, too. The increase has been differential, not all parts expanding equally. For example, the association areas of the cerebral cortex, together with their related subcortical regions, have become greatly enlarged, especially in the parietal region.

This threefold expansion in human brain size has not been accompanied by an equivalent increase in the number of brain cells, for there are only about 25 per cent more cells in the human cerebral cortex than in that of the chimpanzee (this amounts to around 1.4 billion extra cells). However, human brain cells (neurons) are larger, more complex, and spaced further apart than the ape's. Neuroglial cells—cells in close physical and metabolic contact with neurons— also increase in number with increasing brain size. Although it would seem likely that the more complexly branched a neuron is the more interconnections it can make with other neurons and the richer and more varied behavior can be, the specificities of human behavior cannot be explained in terms of any one of these parameters. Much more likely, it is the interaction of these factors with changes in the internal organization and interrelationships of the brain that produce the uniquely human brain output. That organization is important is clear from the fact that human microcephalics (very small-brained people) of various sorts behave in specifically human ways, although they may have brains no larger than those of apes, brains that probably contain fewer cells than those of average healthy gorillas or chimpanzees (Lenneberg, 1964).

These considerations point up the fact that comparing 1 cm^3 of human brain with 1 cm^3 of ape brain is not a comparison of equivalents. It seems equally clear that it is impossible to draw any sort of "rubicon" of brain volume beyond which a creature can be

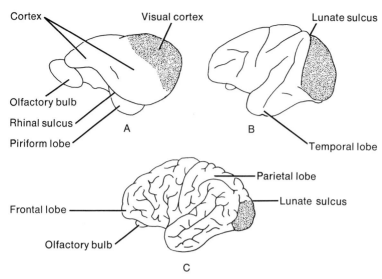

Figure 38 Brains of the tree shrew (*A*), macaque (*B*), and man (*C*), drawn to the same size. In man the cortex is highly folded and the olfactory bulbs are reduced; the parietal lobes are expanded, as are the frontal lobes. Note the more posterior position of the lunate sulcus in C. This "pushing back" is due to parietal expansion in man. (After Campbell, 1966.)

considered human, for the brain did not simply expand but became restructured. Indeed, it can be plausibly argued that the restructuring of the brain probably came first and the expansion (correlated with changes in neuron size, branching, and density) followed later.

The cortex in higher primates can be divided into a number of regions, differentiated in structure and function (Lancaster, 1968). These regions mature at different times, maturation being defined as the acquisition by neurons of a myelin sheath, a kind of insulating coat. The areas that mature first are described as "primordial zones" and consist of the motor cortex, the three primary sensory regions for visual, auditory, and somesthetic (tactile) input, and the limbic system. The motor cortex forms the back of the frontal lobe and

Figure 39 (opposite) Lateral surface (*A*) and medial surface (*B*) of the human brain showing "primordial" areas, those that myelinate (or mature) first. *C* shows the major connections of the visual association cortex in the macaque: *X* to the limbic association area, *Y* to the motor association area, and *Z* to the opposite visual association area. *D* shows the main areas of the human cortex involved in language production: Broca's area, Wernicke's area, and the angular gyrus. The angular gyrus is part of the parietal "superassociation" area receiving input only from other nonlimbic sensory association areas. (After Lancaster, 1968.)

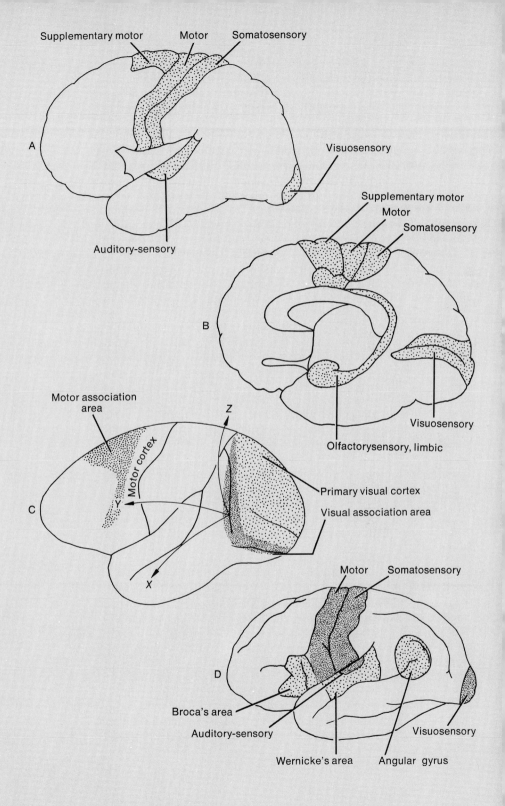

the somesthetic cortex forms the front of the parietal lobe; visual and auditory areas are located in the occipital and temporal lobes, respectively. Deep in the cerebrum, surrounding the brain stem, is the limbic lobe (the term refers to the older—in an evolutionary sense—parts of the cortex and its associated nuclei). The limbic region is closely connected with the lower parts of the brain important in the control behavior vital to reproduction and self-preservation and apparently is intimately involved with all those basically emotional behaviors necessary for species survival.

In the most primitive mammals the limbic region makes up most of the cerebral cortex. In more advanced types the other primordial zones contribute relatively more. In higher primates phylogenetically newer areas of the cortex known as "association areas," which mature later than the primordial areas, have differentially expanded. Some of these association areas receive input from only one of the primordial zones, and it is generally agreed that there is an association area for each primordial zone. The association areas are involved with the integration of behavior. At this point, a little will be said concerning what is known about the functioning—in very broad terms—of various association areas.

At least part of the frontal lobe association areas is involved in drive inhibition. Damage to the human frontal region often produces profound changes in mood and personality. Experimental work with rhesus monkeys has shown that removal of the frontal lobes alters the dominance status of certain animals, principally because previously subordinate animals appear less inhibited, less cowed, in their interactions with animals that are normally more dominant. Other experiments have demonstrated that the neural pathways that mediate these behaviors involve parts of the limbic system as well as areas in the midbrain and brain stem.

The temporal lobes are involved in the storage of auditory and visual information and also in the performance of sequential activities (Stepien et al., 1960). Monkeys can be trained to respond to stimuli given in a specific sequence. For instance, one response is required when a stimulus of type A follows one of type B, whereas another is necessary with the reversed sequence. The stimuli can be visual or auditory—flashes of light or clicks, for example. If certain parts of the temporal lobes are damaged experimentally, this type of sequential behavior becomes difficult or impossible. Once again, limbic interconnections are important in these types of behavior. The temporal lobes and their connections are clearly very important in higher primates, highly emotional and often aggressive animals that live in structured social groups where they are more or less constantly surrounded by their fellows and where continual adjustments in individual behavior have to be made. It is vital in a social context such as this that subtle changes in an animal's emotional state be conveyed to others in the group and that the messages be understood. The social communication of primates is extremely complex, involving postures, gestures, facial expressions, and vocalizations, and utilizes a number of channels—visual, auditory, olfactory, and somesthetic. All these sensory inputs have to be related to limbic (emotional) responses, and they must also be related to each other. A great deal of such social information is conveyed in a sequentially significant manner; the order in which the bits of data come affects the meaning of the whole message.

There are internal connections from each of the three sensory association areas to other parts of the brain. For example, the visual association area in one side of the brain has interconnections with the visual association area in the other side, with the motor association cortex, and with the association region of the limbic system (Lancaster, 1968). Other sensory modalities appear to have the same types of links. Almost all monkey and ape behavior depends upon sensory–limbic interconnections. For example, learning generally requires such associations, whether the behavior is learned by reward or punishment. The temporal lobe is particularly important in forming interconnections between auditory and visual signals and the limbic system.

Another type of connection is said to be at a rudimentary level in apes. This involves the development of a new association area in the parietal region, one having connections only with other association areas and not with primary sensory or motor regions or with the limbic system. It is this part of the brain, the so-called association area of association areas, that has become so greatly expanded in man. Only man is capable of forming, to any great extent, nonlimbic–nonlimbic–limbic associative chains; this is of great importance in any discussion of language or tool making.

There is some debate as to whether or not frontal and temporal lobes have become enlarged "relatively" in the evolution of the hominid brain, and the matter has yet to be resolved (Holloway, 1968). However, the parietal lobe has expanded enormously. It should not be thought, though, that human behavioral specificity lies in the parietal lobe, or indeed that it lies in any particular place. It is clear that the human brain has become reorganized internally, so that although it is still undoubtedly a primate brain its output is qualitatively different from that of other primate brains. We cannot define here precisely how and why human behavior differs from that of our nonhuman relatives. However, it is possible to list some of the more obvious human characteristics.

In discussing the behavior of a mammal with a nervous system as complex as that of higher primates, it is necessary to remember that a great deal of behavior is learned. The old dichotomy between learned and instinctive behavior (and also between environmentally and genetically determined traits) is widely regarded now as being of little or no help. All behaviors are at the most basic level genetically determined, but some of them require enormous amounts of learning and, being largely learned, are subject to considerable variability of expression. Thus, language abilities in man are genetically based; it is almost impossible to suppress language acquisition in a normal human, and the way in which speech develops in children tends to follow a rather regular pattern. However, we do not possess the ability to speak *a* language, but the capacity for the acquisition of language. Individuals within a species generally learn most easily to perform those behaviors that are necessary for survival, and the learning of these behaviors is, in general, in some way pleasurable.

Human behavior is characterized by at least some of the following: the ability to behave cooperatively, to suppress or channel rage and aggression, to sustain motivation or drive over considerable periods of time, to form close affectional ties with other adults of the opposite sex, to make and use tools, and to communicate linguistically. It can be argued, and has been, that these and other attributes are in some way bound up in a behavioral package that developed as the hominids became hunters and gatherers.

Modern hunters live in small bands of around fifty individuals, the band being composed of subunits, the so-called nuclear family, containing an adult male and female united in a formal, legal marriage and their offspring (Lee and DeVore, 1968). To what extent the pair bond between male and female is genetically programmed is unknown. Possibly, it is relatively easy for humans to form close affectional relationships with one other adult of the opposite sex, these ties being sometimes intense and frequently of relatively long duration. But in hunting society other factors are involved, and it can be argued that economic and political factors are what maintain the relationship by reinforcing such biological determinants as there are. There is division of labor between sexes, and it can be argued that the smallest reciprocal economic unit is the pair bond, the hunting and gathering roles complementing each other. Marriage in hunting society is exogamous—mates are found outside the group into which an individual is born—and so marriage also performs a political service in that it spreads the nexus of kinship ties between groups, not just within them. This tends to result in friendly relationships between bands in adjacent hunting ranges, very important during harsh conditions when resources must be shared widely between groups (Washburn and Lancaster, 1968).

Cooperative behavior is strongly developed in hunting society, perhaps more so in males, who need to plan hunting activities. It should be emphasized that this type of behavior in man is no more an instinct than is the pair bond, although it probably does have genetic and neurological foundations that make it relatively easier, given the appropriate circumstances, to learn this type of behavior. Aggressive and dominance behaviors are generally less frequent in hunting society than in societies of other primates. This is probably because of constraints imposed by the economics of the situation, for hunters cannot afford not to be cooperative sharers, and there is no place for dominance hierarchies in such societies. This is not to say that man is not aggressive, or that his aggressive behavior does not have biological roots, but that its control or expression generally depends upon social factors.

The nuclear family contains a number of offspring, often ranging considerably in age. Ties between young and their parents, especially the mother, are very long-lasting and extremely important for the learning of adult skills. Socialization of the young is also profoundly influenced by play with peers. Learning in man differs from that in other primates in duration and intensity and is also greatly faciliated by a system of social rewards; these depend ultimately upon language for their transmission.

Of all the behavioral differences between man and other animals, language and tool making stand out as most distinctive and most important. In general, it can be said that the communication systems of nonhumans convey information about an animal's motivational state and depend directly upon limbic connections, whereas language conveys another type of information and makes statements that do not reflect the internal emotional state (Lancaster, 1968; Reynolds, 1969). This can be tied in with the great enlargement of the human neocortex, particularly those parts that do not have direct connections with the limbic system.

There are many "design features" that distinguish human language from other types of vocal communication, and only a few of them shall be briefly mentioned here (Hockett, 1960). Language is a vocal sequential sign system; more important, the order of the elements within the sequence controls meaning. Thus it is a "relational" system rather than a purely "combinational" one. As we have seen already, higher primate communication also involves sequential signs, some of which may be relational, and it is possible to train monkeys to respond appropriately to relational sequences. However, these abilities are apparently quite limited, and the information conveyed in human language differs from that carried by monkey and ape communication. Part of the human temporal lobe, the so-called Wernicke's area, is important in human language production. As we have noted, this brain area apparently controls sequential signals in nonhuman primates.

Language is an open system; that is, a theoretically infinite number of meaningful messages can be generated. This is possible for a number of reasons, one of which is that language is referential, it can be used to refer to objects (including ideas, qualities, and feelings as well as noun-type objects). It has been suggested that this process of object naming involves the linking of an auditory cue with a cue in some other sensory modality, without directly involving the limbic system. Such cross-modal transfers are easily performed in man but (it is said) can be completed only with great difficulty in nonhuman primates, because the development of the parietal superassociation area in man provides a way of interconnecting sensory modes without utilizing the limbic system. Language also has the property of displacement; it can be used to talk about events in the past or future, or objects out of sight. Once again, emotional state need have nothing to do with such language usage. Finally, language is hierarchic. Meaningful elements (words) are composed of essentially meaningless sounds (this is termed "duality of patterning") and can be further combined into more meaningful groupings. Language can be said to have a nested structure.

Thus, language is the product of a brain that can, among other things, generate and understand relational linguistic codes that are hierarchic, that name objects (in a broad sense), and that permit some degree of displacement. All of these behaviors (although not their linguistic manifestations) are present at least on a rudimentary level in higher primates.

Some effort has been expended in attempts to teach chimpanzees to talk. The Hayeses (Hayes and Hayes, 1955) had tremendous difficulty training Viki, their young female chimpanzee, to utter a very few words, although she possessed good memory, intelligence, and imitative skills. She would vocalize in this way only in a highly specific motivational context and evidently experienced considerable distress in doing so. Chimpanzee brains are not built to generate vocal language. (It should be remembered that some human microcephalics with brains no larger than those of chimpanzees are able to talk.)

More recently, the Gardners (Gardner and Gardner, 1969) have been attempting with some success to teach a young chimpanzee, Washoe, American Sign Language. Washoe has already accumulated a large store of "names," most of which involve visual–visual interconnections, but some of which appear to involve cross-modal transfers. For example, she will use the sign for "dog" upon hearing a dog bark. She has also spontaneously invented new signs and produced sequences of signs, although none of these have yet been relational.

If we assume that the brains of protohominids were capable of producing behavior that, in the appropriate combination, would allow the production of a rudimentary language, what selection pressures might have caused such a change? For want of any better hypothesis, the best suggestion might be that it was the development of hunting as a way of life, for it would certainly seem that planning and cooperation among males, organization and coordination of economic activities in both sexes, and many other activities would require language.

Stone tools first appear in the fossil record of a little more than $2\frac{1}{2}$ million years ago (Howell, personal communication; Leakey, 1970b) and probably were being manufactured even earlier. These tools consist of a number of types, presumably each one serving several functions. Unlike tool modification in chimpanzees, human use of a tool is not limited to one specific function; human tools are more complex—more activities are involved in their manufacture and their form is governed by (arbitrary) norms or rules. Tool making, like language, is a hierarchical activity, and this factor, plus the standardization of tools and the fact that they are the expression of arbitrary imposition of form upon the environment, suggests that the type of brain that was capable of tool making could also generate at least rudimentary language.

Hominids, unlike other primates, are dependent on tools for survival, and tool making is intimately bound up with the hunting way of life. It can be argued that hunting, language, tool making, division of labor between sexes, cooperation, formation of the nuclear family, incest prohibitions, and rules of exogamy are parts of a multifaceted behavioral complex, and, if this is indeed so, then we can trace back at least the rudiments of these behaviors to the time of the earliest stone tools, some $2\frac{1}{2}$ million to $2\frac{3}{4}$ million years ago.

Something can be said concerning the evolution of brain and behavior during the late Pliocene and Pleistocene from a study of natural brain casts and endocranial molds, stone tools, and from associated remains other than the hominids themselves. If we estimate that the hominoids ancestral to hominids weighed around 40 to 50 pounds on average and if we further assume that the relationship between body weight and brain size was the same in these hypothetical protohominids as it is in pygmy chimpanzees, we can give a tentative value of around 300 cm^3 for their mean brain volume. We know nothing of the brain in Miocene *Ramapithecus*, but it is known that late Pliocene *Australopithecus* species had bigger brains than that. The small species, *A. africanus*, weighing around 50 pounds, had an average brain volume of a little less than 450 cm^3 (Holloway, 1970), and the larger forms (*A. boisei*, for example), weighing 150 pounds or more, had brains between 500 and 550

Figure 40 Relationship between brain volume and time in hominid evolution. The capacity of *Ramapithecus* is estimated; the temporal positions for the later species are approximate. (Pilbeam.)

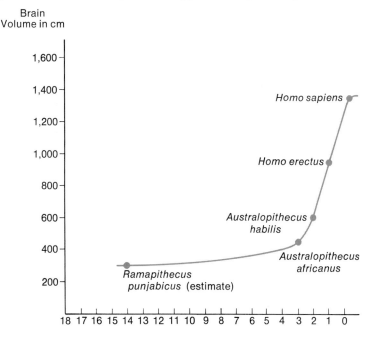

Brain
Volume in cm

cm^3 in volume. In *A. africanus* there is evidence to indicate parietal lobe expansion, and the cerebellum in the more robust species was—at least in external morphology—more like that of man than of the apes (Holloway, personal communication). We known too that at least one of these early species was a tool maker, and both were bipeds. Therefore, all the evidence points to brain reorganization and also to some expansion, due perhaps to internal changes such as increased neuron size and reduced cell density. These forms may well have communicated linguistically with each other and lived in social groups more like those of man than of apes. By the early Pleistocene, *A. habilis,* the "descendant" of *A. africanus,* had a brain volume of around 600 cm^3 or more (Tobias, 1968), although average body size was small, perhaps still no more than 50 or 60 pounds.

After the early Pleistocene, the brain increased rapidly in size, possibly involving some extra differential expansion of the temporal and parietal lobes. Stone tools became more complex, and the incidence of hunting big game also rose. The most reasonable assumption is that social organization was becoming more complex as language grew more efficient and individuals became more intelligent. Presumably, hominid social life was much as it was in later times. The increase in brain size during the last half of the Pleistocene is likely to have been due mostly to an increase in size and complexity of brain cells, rather than to a growth in numbers, and also perhaps to changes in biochemical efficiency. The brain reached its present size at least 100,000 or more years ago, and there were probably few dramatic changes in the biological bases of social organization and social behavior after that time.

SKIN

One important difference between men and apes, the development of which cannot be traced in the fossil record, concerns the structure of the human integument (Montagna and Ellis, 1963). The apes are covered with long, coarse hairs, their skins have few sweat glands, and pigmentation is confined to the deep layers of the skin, the dermis. In contrast, men are much less hairy. Actually, we have at least as many individual hairs as apes do, but they are much shorter and finer, except in a few localized areas such as the head, the armpits, the pubic region, and in males the face and chest. Primate skin glands are of two main types, eccrine and apocrine. Eccrine glands secrete a thin watery fluid, apocrine glands a more viscous substance. The number of eccrine glands has greatly increased in man relative to the apes, and they are distributed all over the skin, although more concentrated in certain areas. Apocrine glands have a more localized distribution and are found in some numbers in the armpits and around the genitalia, for instance.

Finally, in man skin pigment is located in the outer layers of the skin, the epidermis.

Many explanations have been proposed to explain these changes, for we can assume that the very earliest hominids resembled the apes in these features. Human hunting, and gathering, too, often requires more or less continuous activity throughout the day. In the Carnivora, hunting takes place early in the day, at dusk, or at night. The hominid activity pattern requires an efficient means of controlling body temperature, and this we do by secreting watery sweat from the eccrine glands, which is evaporated from the skin and thus cools the body by removing latent heat. This would not be possible were the body covered with dense long hair, hence the necessity for "nakedness." (Strictly speaking we are not naked, but less hirsute—hairy—than the apes.)

Many explanations have been proposed for the pattern of human pigmentation, frequently in attempts to discover why some men are pigmented, the tacit assumption being that the "natural" condition is the white-skinned condition (or, as Professor John Buettner-Janusch would have it, more accurately the "swine-pink" condition). However, because a majority of people living in the tropical and subtropical regions of the Old World are brown or black (taking into account relatively recent population movements) and because earlier hominids, at least until the middle Pleistocene, were living in nontemperate areas, it seems more reasonable to assume that most early hominids were dark-skinned. What might be the selective advantage of epidermal pigmentation? Perhaps the most plausible explanation is that melanin (the pigment mainly responsible for skin color) in the outer layers of the skin protects the all-important eccrine glands from the effects of damaging radiation from the sun (especially ultraviolet light). There is some experimental evidence to support this theory. If all these characteristics do in fact belong to one adaptive complex, then it is quite likely that hunting *Australopithecus* would have been relatively hairless and dark-skinned.

Head hair might serve as protection and might also have evolved as ornamentation. Armpit hair serves an important function in producing human scent, possibly important in sexual behavior (Goodhart, 1960). The apocrine glands of the armpits secrete a sticky fluid that becomes decomposed by bacteria living on the hairs, resulting in the characteristic human body odor. Pubic hair might function as a marker, again in a sexual context. Male facial and body hair is a clear example of a sexual dimorphism, possibly serving to make males appear more ferocious, either to conspecific males or to other animals.

The pattern of human sexual dimorphism is peculiar. For example, canine teeth are the same length in both sexes, and males and females do not differ greatly in body size. The adult human female is unique among primates in the possession of prominent mammary glands surrounded by fat-filled connective tissue. Unlike the mammary glands of other primates, human breasts do not fluctuate greatly in size and after puberty they are permanently enlarged. How can this difference be explained? Nonhuman primate females have an estrus cycle, a period of a few days approximately every 4 weeks when the female actively solicits sexual contact with males. At the same time cyclical changes occur in the region around the genitals involving swelling of the so-called sexual skin, and there are also changes in olfactory stimuli. In contrast, the human female is continuously receptive sexually throughout the cycle, and it has been suggested that breasts have evolved as sexual signs that are permanently present (Andrew, 1964). This is an extremely difficult hypothesis to test. It has been suggested that the evolution of continuous receptivity is bound up with the development of the pair bond, itself a supposed outgrowth of a switch to hunting, and acts as a means of maintaining the bond over a long period. This is an interesting speculation, but one that is probably impossible to verify.

Thus, many of our peculiarly hominid features appear to be intimately bound to the evolution of hunting as a new subsistence pattern. Hominids probably evolved before they became hunters, but the majority of our characteristic behavioral and morphological features seem to have followed the adoption of this unique way of life.

Earliest Hominids

4

The living apes are forest-dwelling forms, and, as far as we can tell, their ancestors were no exception to this rule. The apes of the Oligocene and Miocene lived out their lives, in the main, as arboreal fruit-eating creatures. It is quite likely that some open country was present during this time, but it probably was of limited extent. Open-country habitats became available to primates in Eurasia during the Pliocene; in Africa, it is quite likely that the savannas did not become really extensive until the Pleistocene, and about the only truly open grassland in the Tertiary forests and woodlands of Africa would have been in areas surrounding lakes and rivers that flooded seasonally—places where the climax vegetation would have to be grass and small shrubs.

Toward the end of the Miocene, and certainly by the start of the Pliocene (between 10 million and 14 million years ago), there is evidence that at least two lineages of primates had taken to living at least part of their lives in these newly available open-country niches. Apparently, these forms were feeding in grassland areas.

The first *Gigantopithecus* teeth were discovered by Professor Ralph von Koenigswald in collections of so-called dragon's teeth sold in Chinese drugstores for medicinal purposes. (For references to this section, see Pilbeam, 1970, and Simons and Ettel, 1970.) Von Koenigswald immediately recognized that some of these "dragon's teeth" represented a new primate, one that was almost certainly larger than the gorilla, the largest living primate. Von Koenigswald described these teeth as belonging to a new species, *Gigantopithecus blacki*. Since that time, many more teeth and a number of lower jaws with teeth intact have been located in earlier Pleistocene deposits in Central China. Since the discovery of the first material, two basic views on the position of *G. blacki* have been prevalent: one, that it was an aberrant and peculiar ape; the other, that it

Figure 41 Approximate temporal positions for species of *Dryopithecus, Gigantopithecus,* and *Ramapithecus.* (Pilbeam.)

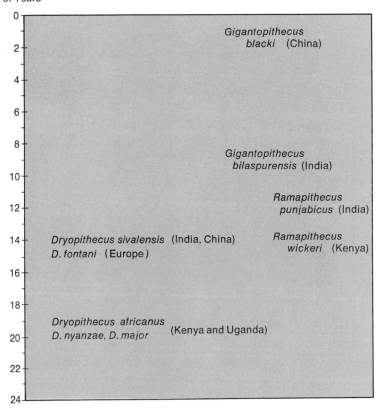

THE ASCENT OF MAN

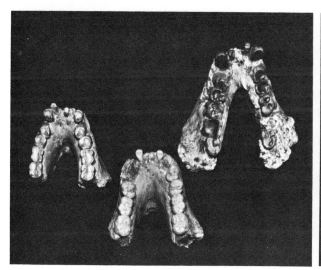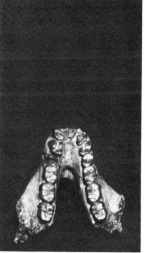

Figure 42 *Gigantopithecus bilaspurensis* (right) and *Gigantopithecus blacki* (left). (Courtesy of E. L. Simons.)

was a giant early hominid. Certainly, some characteristics of *G. blacki* were like those of early hominids: canines did not project beyond the tooth row, incisors were small, and the molars and premolars were relatively huge in proportion to the size of the front teeth. Let us leave the interpretation of Pleistocene *Gigantopithecus* for the moment and turn to a newly discovered specimen from India.

During recent work in the Siwalik Hills of Northwestern India, the foothills of the Himalayas, an expedition under the direction of Professor Elwyn Simons recovered a mandible, with almost all teeth present, of a new species of *Gigantopithecus*. This has recently been named *G. bilaspurensis* (Simons and Chopra, 1969). The mandible came from deposits of middle Pliocene age and is probably 6 million to 9 million years old. The associated animal remains and evidence from sedimentological studies of the rocks suggest that the environment of *G. bilaspurensis* was open-country grassland (Tattersall, 1969). The evidence from China for the habitat of *G. blacki* indicates that similar environments would have been available to that creature, too.

G. bilaspurensis is similar to *G. blacki* in a large number of features (although it is also more primitive), and there is a reasonable probability that the Pliocene form was more or less directly ancestral to *G. blacki*. Both were large creatures, judging from their jaws and teeth, and by analogy with the gorilla—predominantly a ground-living creature—it is most likely that the *Gigantopithecus* species spent most of their time on the ground, foraging for their food in open country as the gelada baboon does in Ethiopia today.

The history of this lineage before the middle Pliocene is not completely clear, but there is some evidence to indicate that

Gigantopithecus evolved from a species of *Dryopithecus*, perhaps *D. indicus*, that is found in Indian deposits of the late Miocene and early Pliocene. If this phylogenetic series is accepted, from *D. indicus* through *G. bilaspurensis* to *G. blacki*, then we are in a position to outline the morphological trends within this lineage, and to interpret them. Through time, animals became larger, with bigger cheek teeth; incisors remained small, however, and canines also became reduced in relative size. Canines had flat surfaces, level with the cheek teeth. Cheek teeth became higher-crowned, with

Figure 43 Ecological shifts within various higher primate lineages. Time scale does not apply to pongids. (After Simons and Ettel, 1970.)

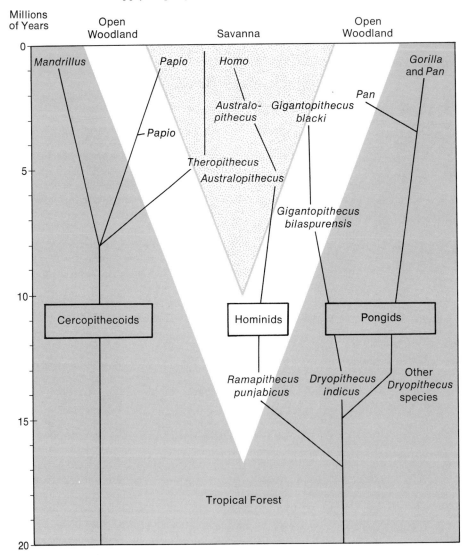

more complex chewing surfaces, and the front lower premolar evolved into a bicuspid tooth as it ceased to shear against the reduced upper canine. During an animal's life, cheek teeth and canines were heavily worn and increasingly compressed against each other as the animal grew older. The mandible itself was massive and thick, rather short from front to back, and heavily buttressed anteriorly, and the ascending ramus rose vertically from the horizontal ramus. Many of these features are those to be found in the living gelada (Jolly, 1970). The small incisors and large cheek teeth suggest that the diet consisted of significant amounts of relatively small morsels such as grass blades, corms, and roots. The large high-crowned cheek teeth, showing considerable wear and crowding, and the massive mandible indicate that *Gigantopithecus* was capable of extremely powerful chewing. Further evidence for this conclusion comes from the short face and the vertical and probably high ascending ramus, indicative of a mechanical system adapted for powerful grinding.

Therefore, all the evidence points to the conclusion that the *Gigantopithecus* species were animals adapted to ground feeding in open country. This is just the type of habitat visualized by most anthropologists for the earliest hominids. Is *Gigantopithecus* a hominid? Probably not, for a number of reasons. First, there were much more plausible human ancestors living at the same time as *Gigantopithecus*. Second, and more important, although *Gigantopithecus* is superficially similar to hominids in tooth proportions and in having nonprojecting canines, the tooth proportions can be explained as adaptations to a particular mode of feeding, and the canines functioned in totally nonhominid ways. Hominid canines are small and incisor-like—sharp cutting teeth. *Gigantopithecus* canines were small, too, but wore to broad flat chewing surfaces—they became, at least functionally, extra cheek teeth.

Therefore, the evidence indicates that *Gigantopithecus* was an aberrant ape, one that became adapted as no living ape is to ground feeding in open country. It is interesting to note that although *Gigantopithecus* had small canines there is no evidence that these species made or utilized tools for defense or offense. How then would such creatures, without large projecting canines and without weapons, defend themselves on the open grassland? Perhaps their large body size was a major factor; they may have had few predators because of this. Possibly, their social organization and display behavior differed in some important ways from those of other primates. These are questions we shall probably never be able to answer.

Perhaps the most important point to emphasize here is that *Gigantopithecus* almost certainly did not make tools yet still had small canines; apparently, in this case, there is no causal relationship between the two. This suggests that the earliest hominids might also have evolved small canines before they invented tool making.

From late Miocene deposits (around 14 million years old) of East Africa and Europe have come remains of another rather peculiar primate, *Oreopithecus bambolii,* a form that, like *Gigantopithecus,* has been said by a number of workers to be an early hominid (Hürzeler, 1958).

Once again, the so-called hominid features adduced for *Oreopithecus* mainly involve dental and facial proportions. For example, *Oreopithecus* has small incisors, relatively short canines, bicuspid front lower premolars, and a short deep face. However, there are a number of other characteristics that indicate that *Oreopithecus* is not related to Hominidae, and this is true particularly for the morphology of the cheek teeth, which are entirely different from those of late Miocene and early Pliocene hominids.

Oreopithecus can therefore almost certainly be ruled out of hominid ancestry and can be seen more plausibly as a representative of an independent hominoid lineage. What is of interest about *Oreopithecus* are its postcranial adaptations (Straus, 1963). *Oreopithecus* had longer arms than legs; it was probably not a knuckle walker, and we can therefore conclude that it was an arm swinger, perhaps rather like the orangutan. Many other adaptations—broad shallow thorax, reduced lumbar region, certain features of the elbow joint—also point to this same locomotor–feeding behavior. In all probability, this complex of features evolved in *Oreopithecus* independently of the similar adaptations found in apes and man.

Some characteristics of the hindlimb, it has been said, indicate that *Oreopithecus* was a biped. For example, the pelvis was broad from side to side and shallow from top to bottom, as in hominids, and the femur apparently had a "carrying angle." However, a number of other features, such as the hip and ankle joints, indicate considerable mobility of the lower limb, adaptations similar to those seen in the orangutan. Possibly *Oreopithecus* was a form that spent its time in the trees mainly as a hanger but was bipedal on its occasional visits to the ground. (This is true of the gibbon, for example.) However, it is perhaps more probable that the so-called bipedal features, particularly in the pelvis, have been misinterpreted. The extinct Malagasy lemur, *Palaeopropithecus,* (Walker, 1967) was an arboreal hanger with extremely long arms and hindlimbs very poorly adapted to bipedal weight bearing, yet had a bowl-shaped shallow pelvis more like that of hominids than any other known primate.

Oreopithecus, like *Gigantopithecus,* is almost certainly not a hominid. Yet we have seen that they both exhibit certain features of similarity with early hominids. It is likely that toward the end of the Tertiary several higher primate lineages were experimenting, so to speak, with new behavior patterns. Probably only one species ultimately survived, that which gave rise to the hominids. It is to this animal that we shall now turn.

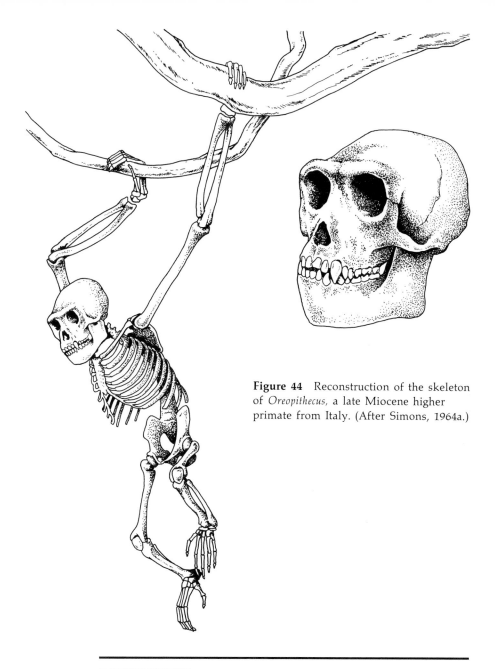

Figure 44 Reconstruction of the skeleton of *Oreopithecus,* a late Miocene higher primate from Italy. (After Simons, 1964a.)

RAMAPITHECUS

In 1934, a Yale graduate student, G. Edward Lewis, described a new primate that he had recovered in the Siwalik Hills of India. He called it *Ramapithecus brevirostris.* Although there were some doubts initially, the age of the original specimen of *Ramapithecus* is now known to be early Pliocene. Other specimens of the same species are also known from late Miocene deposits in the same area, so we can estimate the time spread of *Ramapithecus* in India to have been between about 14 million and 10 million years ago, although these dates are by no means certain. The deposits from which this species comes were laid down, as far as can be judged from the associated fauna and the sedimentological evidence, within a tropical forest environment in an area of low relief through which wide and sluggish rivers meandered. We can assume that rivers of this sort would flood seasonally and that they would therefore be bordered by strips of open grassland. Around 10 million years ago, the forests in Northern India began to give way first to open country and then to grassland.

Since Lewis described *Ramapithecus brevirostris* a number of new specimens have been found; jaws and teeth previously thought to represent other genera have been recognized as belonging to this species (Simons, 1964). Because a specimen first described in 1910 as *Dryopithecus punjabicus* is now known to be a *Ramapithecus,* the species name has had to be changed to *Ramapithecus punjabicus* in accordance with the rules of zoological nomenclature (Pilgrim, 1910; Simons, 1964). Lewis's original *Ramapithecus* specimen was an upper jaw containing some of the cheek teeth, the canine socket, the root of the lateral incisor, and part of the socket for the central incisor. Unfortunately, the other known *Ramapithecus* material provides us with evidence only of upper and lower jaws, part of the lower face, and most of the teeth. No skull parts or postcranial bones have been discovered.

When Lewis first described *Ramapithecus* in 1934, he considered it the most manlike of the dryopithecines. The following year, Aleš Hrdlička, then one of the United States' senior physical anthropologists, wrote a rather polemical paper disagreeing with Lewis (Hrdlička, 1935). This paper had the unfortunate effect of turning attention away from the importance of *Ramapithecus.* Hrdlička argued that *Ramapithecus* did not particularly resemble hominids (that is, modern man), although he did say that it was more similar to man than was *Australopithecus!* (At that time few workers accepted *Australopithecus* as a hominid.)

In 1937, in his Ph.D. thesis, Lewis described *Ramapithecus* as a hominid ancestral to *Australopithecus.* Unfortunately, this study has remained unpublished and has not therefore been available to anthropologists. Until the 1960s, *Ramapithecus* was described in papers and textbooks as interesting, tantalizing, but uninformative.

In 1961, Elwyn Simons began to restudy the *Ramapithecus* material (Simons, 1969). Since 1963 Simons and I have continued research on this interesting primate. Simons became convinced that Lewis was correct in describing *Ramapithecus* as a hominid. He was also able to recognize another maxilla of *Ramapithecus* in the Geological Survey of India collections in Calcutta (Simons, 1963). During our research on the dryopithecines, we noticed that the genus named *Bramapithecus* was hominid-like and that all specimens of *Bramapithecus* were mandibles, whereas all undoubted *Ramapithecus* were

Figure 45 Side view (*A*) and occlusal view (*B*) of an upper jaw fragment of *Ramapithecus punjabicus*. (Courtesy of E. L. Simons.)

A

B

maxillae. It was decided that it was highly likely that these manlike lower and upper jaws in fact belonged to the same animal, *Ramapithecus punjabicus* (Simons, 1964).

The lower face of *R. punjabicus* is known and indicates that compared with a living primate of approximately the same size, such as a pygmy chimpanzee, the face was relatively deep from top to bottom and short from front to back. The zygomatic arch is situated above the first molar, indicating that the origin of the masseter muscle was relatively forward, thus increasing the lever arm of this important chewing muscle (which would raise its power). The lower jaw is thick and shallow posteriorly, deeper anteriorly, with a vertical chin region buttressed internally by a relatively large inferior torus. The ascending ramus rose vertically from the horizontal ramus, beginning its ascent further forward than is the case in the pygmy chimpanzee. All the features indicate a form with a short, deep, well-buttressed face and jaws; this complex of adaptations suggests powerful side-to-side chewing as in the gelada and *Gigantopithecus*.

Crowns, roots, or sockets of all the upper teeth of *R. punjabicus* are preserved. The incisors were very small teeth relative to the posterior dentition, judging from their roots (Simons, 1969b). Compared with the frugivorous pygmy chimpanzee, *R. punjabicus* incisors were not only much smaller but also more vertically oriented. Canine crowns are unfortunately not known for Indian *Ramapithecus*, but the root sockets are preserved and show that the crowns would have been broad from side to side and shortened mesiodistally (in the long axis of the rest of the tooth row). This suggests very strongly that the canine crown of *R. punjabicus* was, or was becoming, an incisor-like tooth with a broadened cutting edge, as in all later hominids. Thus it seems likely that the incisors formed a relatively vertically oriented almost continuous (there being probably only small gaps between incisors and canines) cutting edge, set in a short face. This implies an adaptation for powerful anterior slicing, since the load arm of the lever system was reduced.

The cheek teeth had flat, broad chewing surfaces, in contrast to the constricted occlusal areas of the Tertiary apes (Pilbeam, 1968). Such teeth are better adapted for side-to-side shearing than for the more vertical movements inferred for dryopithecines. The enamel on the teeth also seems to have been thicker than in apes. During life, the cheek teeth became closely packed one against the other as chewing forces caused newly erupted teeth to shift forward. In a number of ways the molars and premolars of *Ramapithecus* were intermediate between those of *Dryopithecus* and *Australopithecus*; the upper dentition seems to have functioned more like that of later hominids than like that of the apes.

The lower dentition is somewhat less well known than the upper. Although crowns of incisors and canines are not preserved, there is evidence to suggest that these teeth were small and vertically implanted. The roots only of the front lower premolar are known but suggest that this tooth was not elongated and sectorial as in apes. This conclusion is supported by the morphology of the first upper premolar, which was not adapted for shearing against a blade-like lower premolar (Pilbeam, 1969b). The last lower premolar and the molars all have broad, flat, rounded crowns that are packed closely together. The gradient of tooth wear from first to second to third molars is very marked, suggesting that coarse food was being chewed, so that one tooth would become heavily worn before the next one erupted. This might also imply that eruption was delayed. If this were the case, we could conclude that the infancy and adolescence of *Ramapithecus* was already becoming longer than in the apes.

Taken as a whole, the evidence to be obtained from the *Ramapithecus* material points to an animal with a facial–dental complex of characteristics similar to or foreshadowing those of later hominids, both morphologically and functionally. The relative proportions of front and back teeth, as well as many other features, suggest that *R. punjabicus* was, or was becoming, a ground feeder eating a diet including numerous small, tough morsels. Although *Ramapithecus* remains come from areas that were forested (Tattersall, 1969), as noted already open areas were probably present around rivers and lakes. It is therefore possible that *R. punjabicus* was a species in the process of adapting to living, or at least feeding, in these open areas.

Ramapithecus had small anterior teeth; probably the canines projected hardly at all beyond the other teeth—although we cannot be sure of this—and yet this species almost certainly did not make tools. It may well have utilized objects to obtain food and in display behavior, as does the chimpanzee today, yet this cannot be interpreted as the dependence on tools for survival that characterizes the later hominids. As with *Gigantopithecus*, we have further evidence for the view that there is no necessary causal relationship between tools and small canines.

In 1961, Professor Louis Leakey's excavators recovered some pieces of upper jaw and one lower tooth of a form resembling *R. punjabicus* from Fort Ternan in Kenya (Leakey, 1962). This site has been well dated radiometrically at 14 million years and is therefore as old as the oldest Indian *Ramapithecus*-bearing deposits (Bishop *et al.*, 1969). Like the Indian sites, Fort Ternan was probably also once forested.

Leakey described his new find as *Kenyapithecus wickeri*, thus implying that it was generically different from *Ramapithecus*. Both Simons and I believe that the differences between the two forms are insufficient to warrant generic distinction (Simons and Pilbeam, 1965). In fact, they may not even be separate species, although the best course is probably to refer the Kenyan form to the species *R. wickeri*. In almost all features, it resembles Indian *Ramapithecus*. Of particular interest is the canine crown, unknown in *R. punjabicus*. This is very small, although still morphologically apelike, and may well prove to be one feature of distinction from *R. punjabicus*.

The associated animal remains at Fort Ternan and in the Siwalik *Ramapithecus* deposits are quite similar (Simons, 1969b), suggesting that at the end of the Miocene the region comprising East Africa, Arabia, and India formed a large and relatively continuous faunal zone that was largely forested. *Ramapithecus* probably evolved around 15 million years ago, perhaps in Africa, but spread relatively rapidly thereafter throughout at least this African–Asian faunal zone. It seems to have been predominantly a forest creature, although it was probably a ground feeder utilizing feeding opportunities in the open grasslands around watercourses and lakes and at the forest margin.

Was *Ramapithecus* a hominid? The morphological evidence and functional conclusions drawn therefrom suggest that it was an animal with the dental and facial organization characteristic of later hominids, in developed or incipient form (Pilbeam, 1968). On paleontological evidence we can therefore place it on the line leading to *Australopithecus* and *Homo*, after this lineage diverged from the other hominoids. Whether or not *Ramapithecus* is to be described as a hominid depends on where along the line leading to man the hominid–pongid boundary is to be drawn. Of course, this boundary is arbitrary. Hominids are distinguished from pongids paleontologically by three sets of characteristics: cranial features, dentition,

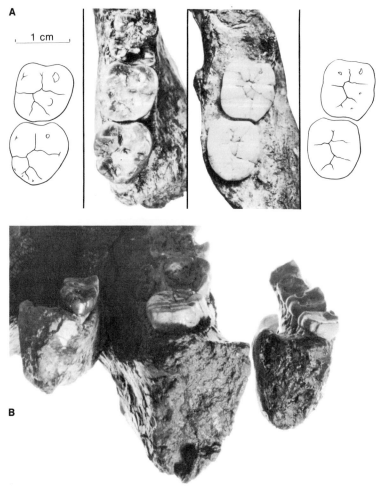

Figure 46 *A*: Comparison of (reversed) *Ramapithecus* mandible (left) with that of a South African Pleistocene hominid. *B*: Comparison of *Ramapithecus* mandibles (left and right) with those of *Australopithecus boisei*. (Courtesy of E. L. Simons.)

and postcranial skeleton. Only the dentition is known for *Ramapithecus*, and this is more hominid- than pongid-like. For this reason, at present I would classify *Ramapithecus* as a hominid.

Two biochemists at the University of California at Berkeley, Vincent Sarich and Allan Wilson (1968), have recently stated their belief that hominids diverged from apes only 4 million or 5 million years ago. This would imply that the hominid resemblances of *Ramapithecus* are parallelisms and not proof that *Ramapithecus* was an early or ancestral hominid. Sarich and Wilson's theory depends upon a comparison of serum albumins of a variety or primates. They believe that albumin in primates has evolved at a constant rate and therefore that the difference or "distance" between the albumins of any pair of primates can be used to calculate the time since their separation. For example, apes and men prove to be much more similar than apes and monkeys, and Sarich and Wilson conclude that this indicates a much more recent common ancestry.

These biochemical conclusions concerning hominid origins conflict with the paleontological evidence. One explanation may be that the fossil evidence has been incorrectly interpreted. However, we know now that *Australopithecus* extended back in time to at least 5 million or more years ago and that *Ramapithecus* was, in many ways, very similar to *Australopithecus.* It has also been pointed out that the biochemical evidence can be made to fit the paleontological data, granted certain assumptions about the relationship of time and "distance" (Uzzell and Pilbeam, 1971). One possible explanation for the biochemical similarities between apes and man may be that the generation length of hominoids is considerably greater than that of other primates. Random changes in DNA appearing during the maturation of sex cells will become spread and fixed in a population at a rate proportional to the generation length. These random changes (this applies only to characteristics subject to little or no selection) will accumulate more slowly, and for a given period of geological time there will be less difference between animals with long generation times than between those with shorter generations. Recent work at the Carnegie Institute on DNA hybridization has demonstrated a reassuring compatibility between dates of divergence estimated from the fossils and the amount of DNA differences between primates, taking into account differences in generation time (Kohne, 1970). Although speculative, at least for the present I would rather accept the fossil evidence that suggests that *Ramapithecus* is the earliest known hominid, a form in the process of adapting to ground feeding in open areas within the Tertiary forests of Asia and Africa.

By utilizing recent research concerned with the hypothetical locomotor behavior of man's prebipedal ancestors, it is possible to say something about locomotion in forest forms such as *Ramapithecus,* even though no postcranial bones of this genus have been discovered. Professor Sherwood Washburn (1968) has suggested that man's prebipedal ancestors were knuckle walkers like the living African apes. Professor Russell Tuttle has extensively studied the locomotor adaptations of the chimpanzee and gorilla and has recently concluded that following "dissections of numerous ape and human hands, there are no features in the bones, ligaments, or muscles of the latter that give evidence for a history of knuckle-walking" (Tuttle, 1969:460).

Tuttle's colleague at the University of Chicago, Professor Charles Oxnard, has also been working on this problem, but from a different viewpoint (Oxnard, 1969a,b). Oxnard has been studying the shoulder girdles of a whole range of primates. Following a series of dissections, Oxnard and his colleagues defined a number of ratios and indexes on the scapula, clavicle, and humerus that are thought to have functional significance. These values were then combined and analyzed in a sophisticated series of multivariate statistical studies. Such analyses enabled Oxnard to differentiate clearly between the locomotor behavior of the various primate genera. Man is unique in the structure of his shoulder girdle, as might be expected, but is closest to arboreal forms rather than to terrestrial species such as baboons, chimpanzees, or gorillas.

Oxnard has further been able to establish which primate is closest to man in terms of the minimum number of functionally significant changes needed to produce a shoulder girdle of human type. This form is the orangutan, and Oxnard concludes that the ancestor of man used its shoulders and arms in ways reminiscent of the orangutan; that is, man's prebipedal ancestors were likely to have been large-bodied arboreal forms capable of hanging and swinging by their arms. The evolution of the human shoulder could, according to Oxnard, have simply involved "the *loss*... of the single function of raising the arm above the head for purposes of suspension of the body weight during foraging and locomotion" (Oxnard, 1969a:326). What little is known of the shoulder girdle of *Australopithecus africanus* does not contradict this conclusion (Oxnard, 1969a).

We know that by the end of the Pliocene the hominids were fully bipedal forms although their upper limbs still retained evidence of previous arm-swinging behavior (Oxnard, 1969a). The body size of *Ramapithecus* can be estimated at around 40 to 80 pounds, similar to that of the pygmy chimpanzee. On the basis

Figure 47 Tentative relative times of development of the major adaptive complexes in Hominidae. (Pilbeam.)

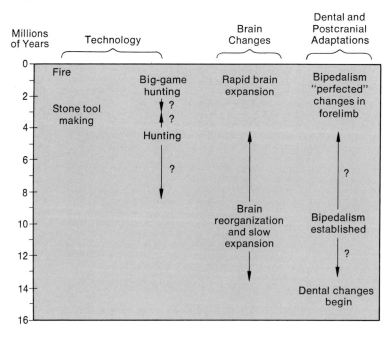

of Oxnard's studies, we can therefore hypothesize that *Ramapithecus* was a predominantly forest-living, arboreal animal, capable of arm swinging and suspension in the trees, yet coming to the ground for food. (However, the possibility should be borne in mind that *Ramapithecus* was a knuckle walker.) In the open foraging areas it was probably bipedal. This phase of hominid locomotor evolution could well have lasted for millions of years, until bipedal behaviors were sufficiently well developed for the hominids to become more or less habitual ground-dwelling bipeds. It should be stressed that this hypothesis is extremely speculative, and we badly need to find fossil limb bones to strengthen or invalidate it.

HOMINID ORIGINS

Until recently it was thought that no fossil evidence relevant to hominid origins had been discovered; however, it now seems quite likely that *Ramapithecus* was an early hominid. Before the recognition of *Ramapithecus* it was generally believed that the first hominids (recognizable by their dental and postcranial adaptations) differentiated during the Pliocene in the open savannas of the Old World. These first hominids, it was thought, became tool users and hunters; their adoption of tools as weapons resulted in the diminution of male canines.

If *Ramapithecus* was indeed a hominid, then man's earliest ancestors differentiated not in open savannas but within and at the edge of forests, utilizing mainly vegetable foods gathered on the ground in the open areas around rivers and lakes. There is no evidence to suggest that *Ramapithecus* was a hunter or tool user, at least not any more so than the chimpanzee. It should be stressed once again that this interpretation is far from certain, yet it is based on concrete evidence, unlike previous hypotheses.

The earliest hominids presumably evolved toward the end of the Miocene from forest-dwelling apes that were relatively large-bodied arm swingers or knuckle walkers. It is possible that the hominids had evolved many of their distinctive dental adaptations before they became habitual bipeds. Initially, they may not have indulged in bipedal behaviors any more than do the living apes. Indeed, the period during which they retained arboreal hanging and swinging or knuckle-walking capabilities may well have lasted several million years. However, we can assume that during the Pliocene hominid bipedal behavior became much more efficient, perhaps as our early ancestors were developing into hunters. It is to this later stage of hominid evolution that we now turn.

5 South African Early Hominids

Since 1924 a large number of fossil hominids now known to range in age from middle Pliocene to early Pleistocene (between 5 million and $1\frac{1}{2}$ million years old) have been recovered in various parts of Africa. These hominids, belonging to the genus *Australopithecus,* have been variously described as "australopithecines," "early Pleistocene hominids," "ape men," "man apes," and so on. We shall refer to them collectively as Plio-Pleistocene hominids.

The first of the forms was discovered in 1924 at Taung in Cape Province, South Africa, during mining operations for limestone (Dart, 1925). The specimen is a child's skull, largely complete, with a natural brain cast formed from limestone. The dentition is mostly milk teeth, although the first permanent molars had erupted. The skull came into the possession of Raymond Dart, then a young professor of anatomy at the University of the Witwatersrand in Johannesburg. Dart named it *Australopithecus africanus* in 1925. He thought that it showed a curious blend of human and ape features and placed it in a new family intermediate between Hominidae and Pongidae. Later he decided that *A. africanus* was indeed a hominid close to the ancestry of man.

These conclusions did not sit at all well with European—particularly English—anthropologists. For one thing, Dart was thought to be inexperienced. Also, *Australopithecus* combined a small brain with a manlike dentition, and this conflicted with the general "consensus" view that the earliest hominids were characterized by large brains and apelike jaws and teeth (a theory based upon the infamous Piltdown skull, known now to be a fake). Thus Dart's skull simply did not fit in with preconceived notions (or hypotheses, as they are now called in paleoanthropology). Finally, a number of workers seem to have been somewhat offended because Dart did not consult them about the skull. So, the status of *Australopithecus* remained equivocal.

At least one paleontologist, Dr. Robert Broom, was convinced that Dart was right, and he began searching near Pretoria for more early hominid sites. In 1936, he recovered an adult hominid skull from Sterkfontein in the Transvaal (Broom, 1936). This he described originally as *Australopithecus transvaalensis,* but in the following year he transferred it to a new genus, *Plesianthropus.* Since then a large

Figure 48 Side view of the infant skull from Taung, the type specimen of *Australopithecus africanus.* (Courtesy of P. V. Tobias and Alun R. Hughes.)

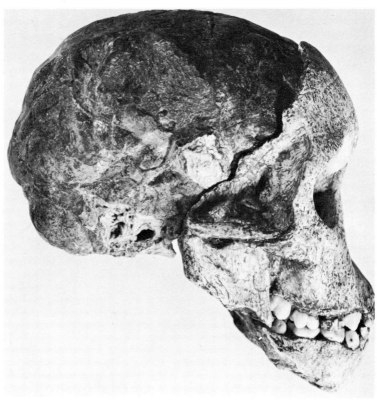

number of specimens—mainly teeth and jaws but including skulls and postcranials—have been recovered from Sterkfontein, and digging has recently started there again.

Two years later, Broom located another site, Kromdraai, close to Sterkfontein, from which he obtained a fragmentary skull with much of the dentition and several postcranial bones. A few more specimens were recovered in subsequent years. Broom called the

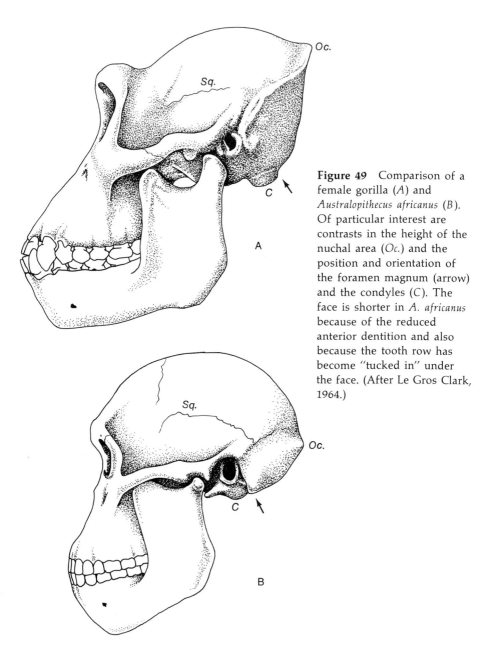

Figure 49 Comparison of a female gorilla (A) and *Australopithecus africanus* (B). Of particular interest are contrasts in the height of the nuchal area (*Oc.*) and the position and orientation of the foramen magnum (arrow) and the condyles (C). The face is shorter in *A. africanus* because of the reduced anterior dentition and also because the tooth row has become "tucked in" under the face. (After Le Gros Clark, 1964.)

Figure 50 Major *Australopithecus* sites in South and East Africa. (After Tobias, 1967.)

new form *Paranthropus robustus* (Broom, 1938). In 1948, Broom found yet another site in the same area, at Swartkrans (Broom and Robinson, 1949). The first specimens he described as *Paranthropus crassidens*. Since 1948 many more specimens have been recovered from Swartkrans, and excavation is in progress there now under the direction of Dr. C. K. Brain (1967a) of the Transvaal Museum.

Throughout this period of time, debate continued on the status of *Australopithecus* and *Paranthropus,* at times reaching quite acrimonious levels. Sir W. E. Le Gros Clark, then professor of anatomy at Oxford, visited South Africa during the 1940s. At first skeptical about the hominid status of the South African primates, after studying them he became convinced that they were indeed hominids, and his "conversion" did much to widen their acceptance as human relatives (Le Gros Clark, 1947); very few workers now doubt their hominid status. More hominids were discovered at a fifth site, Makapansgat, in the Transvaal, by Dart in 1947, and even more specimens were added during the following years. Dart (1948) described this material as *Australopithecus prometheus.*

Thus between 1925 and 1948 remains of early South African hominids from five sites were described and placed in three genera and five species. Work on the faunas associated with the hominid remains suggested a relative chronology for these sites. Unfortunately, no material suitable for absolute dating has yet been recovered, although work on this problem is in progress. It is generally agreed that Sterkfontein is older than Swartkrans, perhaps considerably older (Cooke, 1963; Howell,, 1955). (These are the sites with the largest samples of both hominids and other animal remains and therefore the the easiest to date.) Makapansgat is thought to be somewhat older than Sterkfontein; Taung may be within the time span covered by these two sites; Kromdraai is generally considered somewhat younger than Swartkrans. Sterkfontein is probably older than 2 million years, judging from comparisons with the well-dated East African sequence; it may be as old as $2\frac{1}{2}$ or more million years (Cooke, personal communication). Makapansgat is $2\frac{1}{2}$ to 3 million years old, and Swartkrans and Kromdraai probably belong in the 2 million to $2\frac{1}{2}$ million year period. It should be stressed that these ages are all extremely tentative.

An attempt to estimate the climate at each of these sites during the period of deposition has been made in an elegant study by Dr. Brain (1958). Brain discovered that certain types of soils could be analyzed in such a way that the local rainfall at their time of deposition could be estimated. The deposits from which the South African hominids come were once caves in which bone, soil, and other debris accumulated. This material eventually became consolidated to form breccia. Erosion since the time of deposition has stripped away the roofs of the caves to expose their contents. By analyzing the breccia, Brain has been able to estimate for each site the local rainfall when the hominids were being fossilized. He concludes that the climate at Sterkfontein and Swartkrans was a little drier than it is today in that area, whereas that at Makapansgat was considerably drier; conversely, Kromdraai was wetter than today.

The habitat in the areas surrounding these sites today is open country, with occasional woodland, particularly along streams. It is possible to estimate very tentatively the vegetational pattern in these areas with lowered and raised rainfall (Cooke, 1963). Although the distribution of grassland and woodland would have altered somewhat, the habitat would probably have remained relatively open whatever the changes in rainfall. Thus we can conclude that the early hominids were living in an open habitat, and certainly not a forested or even woodland one.

In 1954, Professor John Robinson produced a revised classification of the South African early hominids in which he allocated them to two genera and species, *Australopithecus africanus* (sampled from Taung, Sterkfontein, and Makapansgat) and *Paranthropus robustus* (Kromdraai and Swartkrans). Thus *A. africanus* is older than *P. robustus* in the South African sequence. Robinson believes that these species are not sampled from a single evolving lineage, but rather represent two phyletic sequences separate for a considerable period of time. The two species differ in a number of characteristics; perhaps most importantly the molars and premolars of *P. robustus* are considerably larger than those of *A. africanus*, whereas the incisors and canines are much the same size in the two forms. Robinson has proposed that the differences in cheek-tooth size and dental proportions are due to the fact that *Paranthropus* was a strictly vegetarian species and *Australopithecus* was more omnivorous. This theory has been termed by Professor Phillip Tobias (1967) "the dietary hypothesis."

Tobias (1967), following his detailed work on an East African early Pleistocene hominid called *Zinjanthropus boisei*, has offered an alternative classification. He believes that *A. africanus* and *P. robustus* are best classified into a single genus, *Australopithecus*. He further concludes that the evidence adduced to support the dietary hypothesis is insufficient, and he assumes that the two species of *Australopithecus* were similar in feeding behavior.

The following account will discuss the material from each site, its relative age, and its taxonomic position. As far as possible, attention will be drawn to the still considerable interpretative problems—of dating, ecology, functional morphology, and classification—that must be solved before we can begin to understand what happened and why.

TAUNG

Unfortunately, the site at Taung has been destroyed by mining operations. Few associated animal remains were found with the first *Australopithecus* specimen, and so the relative stratigraphic position of the hominid is hard to assess. However, the evidence does seem to weigh rather more in favor of an earlier than a later age. No stone tools have been found at Taung, and it is not now possible to make any definite statements about climate at the time of deposition.

The specimen, the type of *Australopithecus africanus*, consists of much of the front part of a child's skull together with most of the lower jaw. All milk teeth are intact along with the first permanent molars. Most of the brain has been preserved as a natural limestone endocast. Dart (1925) believed that certain facial features of *Australopithecus* were closer to those of hominids than to pongids. Some 5 years after the skull's discovery, Dart managed to separate the lower and upper jaws, thus exposing the occlusal surfaces of the teeth. He recognized immediately (Dart, 1934) that *Australopithecus* had milk and permanent teeth that were essentially similar to man's and not to those of apes; however, they differed in being much larger than the teeth of modern man. Of course, it could be argued that large size is an apelike feature, but this means very little and certainly would not imply that *A. africanus* was an ape. All the chewing teeth were large, and the front milk molar was broad with several cusps, not single-cusped as in most apes. The teeth were also worn flat, and this combination of characteristics suggests that *Australopithecus* was capable of powerful chewing involving transverse and rotatory movements.

The brain cast of the Taung child is almost complete. As far as can be estimated, the brain volume had adulthood been reached (the infant died at around 6 years of age) would have been about 440 cm^3 (Holloway, 1970). This value lies almost midway between the mean brain volumes of chimpanzees and gorillas; as far as can be estimated, *A. africanus* was a considerably smaller animal than even the chimpanzee.

The external morphology of the Taung brain, as far as it can be discerned, is of considerable interest. The lunate sulcus, which marks the boundary of the occipital and parietal lobes, is situated farther back than in apes, resembling more closely the human condition; this was pointed out originally by Dart (1925) and has since been confirmed by Holloway (personal communication). It indicates differential expansion of the parietal association areas, a characteristically human trait. As discussed previously, in man the parietal association areas appear to be involved in language pro-

Figure 51 Occlusal view of mandible of infant *Australopithecus africanus* from Taung. (Courtesy of P. V. Tobias and Alun R. Hughes.)

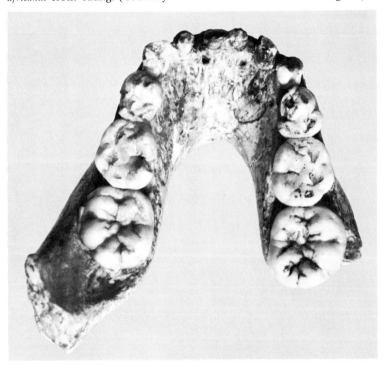

duction and probably other kinds of behavior utilizing symbols, too. Hence *Australopithecus* may well have been capable of rudimentary speech and was probably on the way to becoming a "cultural" animal in ways in which men are, and apes are not.

STERKFONTEIN

The situation at Sterkfontein is much more satisfactory than at Taung, for there are many more hominids known as well as a good representative series of associated mammals. As previously noted, the general consensus among paleontologists seems now to favor an early age (late Pliocene) for Sterkfontein. In earlier descriptions of this site, it was stated that a portion of the deposit, known as the Extension Site (Robinson, 1961), was younger in age than the original, or Type Site. Stone tools were found at the Extension Site and were thought not to occur in the supposed older deposits. Hominids were found in both areas. However, recent work indicates that the deposits may be of essentially similar age (Tobias and Hughes, 1969). The tools have been said to resemble those from deposits in East Africa that can be dated at around $1\frac{1}{4}$ million to $1\frac{3}{4}$ million years old. It is possible that the Sterkfontein tools are much older than that, and certainly the Type Site apparently indicates an earlier rather than a later date. Provisionally then we can assign Sterkfontein to the period between 3 million and 2 million years ago (say, around $2\frac{3}{4}$ million years). This would be latest Pliocene, if the Pliocene-Pleistocene boundary falls at just under 2 million years ago. Almost all hominids are from the Type Site.

It is difficult to evaluate with accuracy the taxonomic status of the Sterkfontein sample (which does seem to represent one species) relative to the Taung child. However, at present there is no reason to assume that together they do not represent one species, and therefore the Sterkfontein specimens will be treated as *A. africanus.*

Several more or less complete skulls of *A. africanus* are known. The brain case is rounded with a relatively well-developed forehead (Robinson, 1964). Moderate brow ridges surmount a rather projecting face, which is nevertheless quite deep. Superficially, the combination of large face and small brain makes *A. africanus* appear somewhat apelike. But this is simply due to the relative sizes of brain and face (the latter being related to the size and mechanical effectiveness of the dentition). In a number of quite detailed features that need not be listed here *Australopithecus africanus* resembles the later Hominidae.

The teeth of *A. africanus* are well represented (Robinson, 1956) and exhibit typically hominid features of morphology and organization. Incisors are spatulate and set vertically in the jaws, and the canines are short and incisor-like teeth that project hardly at all above the other teeth. There are no gaps between incisors and

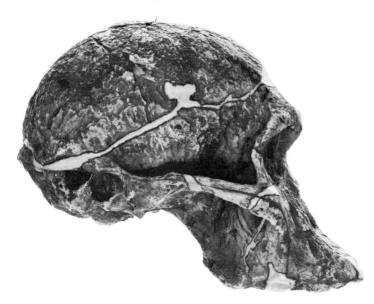

Figure 52 Side view of a skull (cast) of *Australopithecus africanus* from Sterkfontein (Courtesy of Wenner-Gren Foundation.)

canines, and thus the front teeth form a continuous vertically oriented cutting battery. The entire dental arcade is parabolic as in modern man, the cheek teeth on either side diverging toward the back. This is because the face is short and the canines are small, and also because the efficiency of grinding and slicing is increased by this arrangement.

Cheek teeth of *A. africanus* are larger than in later hominids and are set in quite robust upper and lower jaws. Morphologically they are very similar to teeth of *Homo,* having rounded cusps composed of thick enamel. This allows the teeth to wear flat before the enamel is perforated. Flat tooth wear indicates that lateral and rotatory chewing movements were important. The molars show marked differential wear (Robinson, 1954); the first lost its enamel before the second was heavily worn, and so on. This is a reflection of two factors: first, heavy tooth use and, second, delayed eruption of the teeth—implying that *A. africanus* took longer to attain sexual maturity than do the living apes. The thickened enamel was probably an adaptation to the longer life period of hominids.

Although superficially similar to those of apes, the skull and dentition of *A. africanus* exhibit many characteristics that suggest that this species could have been ancestral to, or closely related to the ancestry of, the genus *Homo.* The main determinants of skull form are the masticatory apparatus and the brain. In the known *A. africanus* skulls, areas for insertion on the skull of masticatory and neck muscles are weakly developed, and so crests and ridges are not particularly large. (The possibility should be noted that all or most known *A. africanus* skulls may be female; these would be less robust than those of males.) Thus, the shape of the brain case follows quite faithfully the external contours of the brain (Tobias, 1967). The face is relatively deep and short compared with those of apes. The deepening was an adaptation to a herbivorous diet, permitting more controlled and powerful transverse and rotatory movements of the lower jaws (Crompton and Hiiemäe, 1969). The shortening of the face had a similar effect; inasmuch as the vertically directed incisors and canines are set in a short face, more power could be exerted in slicing and cutting because the load arm of the lever system was reduced.

What can be said about masticatory movements? Large and flat-crowned cheek teeth, thick mandibles, thickened tooth enamel, the pattern of tooth wear, and the short, deep face suggest very powerful chewing; the diet was probably basically vegetarian. The anterior slicing teeth may have been utilized in preparing vegetable food or meat. It is probable that meat was eaten, judging from the likely association of tools with *A. africanus*. Living hunter–gatherers consume perhaps two thirds or more of their food in the form of vegetable matter (Lee and DeVore, 1968), and this was probably the case in *A. africanus*. *A. africanus* can probably best be described as omnivorous and may well have been an eclectic feeder like the living savanna baboons of Africa.

There is a possibility that all the Sterkfontein skulls (there are four) for which brain volumes—strictly speaking, skull volumes—can be estimated are female. The estimates obtained by Holloway (1970) range from 428 to 485 cm^3, giving a mean of 444 cm^3, not materially different from the Taung estimate. It is possible to estimate the size of jaws that would be associated with these small skulls. There are present at Sterkfontein much larger jaws than these, and so it can be inferred that males of *A. africanus* may well have had somewhat larger brains and probably had skull crests due to their larger masticatory muscles.

The stone tools from Sterkfontein Extension Site (Robinson, 1961) include various types of cutting and pounding tools that would have been utilized in the preparation of both vegetable food and meat. Robinson (1959) has described one Type Site bone tool that was probably utilized in preparing skins, possibly for use as clothes, blankets, or as crude shelters. Although the cave at Sterkfontein was almost certainly not itself used as a living site, the hominids may well have been occupying natural rock shelters nearby. Their remains, together with those of tools and other mammals, would have been washed into the cave by rains and surface runoff. Many of the associated animals were no doubt collected by the hominids for food.

A small but nevertheless important series of postcranial bones has been collected at Sterkfontein. These have been studied by a number of workers, including Robinson (personal communication), Le Gros Clark (1964), Oxnard (1969a, b), Dr. Bernard Campbell (personal communication), and Professor Adrienne Zihlman (1967). The upper limb is known from a fragment of scapula and various parts of the humerus. Although these pieces do show some human characteristics, they do differ from man in a number of ways.

According to Oxnard, they suggest an animal that was capable, or whose ancestors were capable, of suspending the body from fully extended arms. *A. africanus* lived in open country and, as we shall see, was a well-adapted biped, so it is unlikely that suspensory posturing formed a major component of its locomotor repertoire, although it probably retained the capacity to climb efficiently in trees. Relative to trunk length, the human forelimb is somewhat shorter (about 10 per cent on average) than that of the chimpanzee. If it is assumed that the earliest hominids were relatively long-armed hangers, then it is quite likely that *Australopithecus africanus*

Figure 53 Side view of vertebral column and pelvis of *Australopithecus africanus* from Sterkfontein. Figure 54 is a front view of the same pelvis. (Courtesy of J. T. Robinson.)

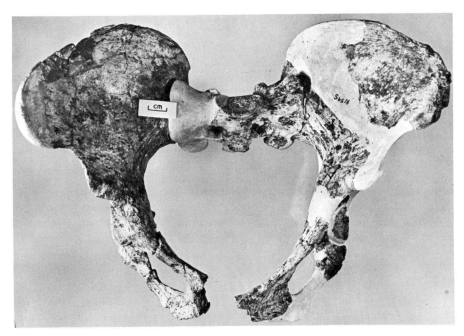

Figure 54   Pelvis of *Australopithecus africanus* from Sterkfontein. (Courtesy of J. T. Robinson.)

had relatively somewhat longer arms than modern man. The Sterkfontein humerus has an estimated length of around 300 mm, a little longer than in chimpanzees and shorter than in male humans. The specimen is said to be from a large male but nevertheless suggests that the arms of *A. africanus* were relatively longer than are those of man.

The trunk and lower limb are represented by most of a vertebral column, pelvis, and upper femur of a small, probably female individual (Robinson, 1963); by parts of another pelvis and femur; and by two pieces of distal femur. Some estimate of body weight can be made from the associated vertebrae and pelvis; it was probably around 50 pounds or less. If this represents a small or female individual, then we might assume a range of body weights for *A. africanus* of between 40 and 70 pounds or so, approximately the same as the pygmy chimpanzee. (However, note that the jaws and teeth of *A. africanus* were considerably more massive than are those of the pygmy chimpanzee, a forest-dwelling frugivorous form.)

The vertebral column has six lumbar vertebrae (only about 3 per cent of humans have six, the modal number being five). The apes never have six lumbars (Schultz, 1968), the knuckle walkers in particular having reduced that region of the vertebral column to three or four vertebrae. This also implies that *Australopithecus* probably did not pass through a knuckle-walking phase. The sacrum is smaller than in man, and this may be true when corrections have been made for body size. The lumbar vertebrae and sacrum form a distinct lumbar curve, characteristic of man and not seen in nonbipeds. The pelvis is broad and bowl-shaped as in man, shortened from top to bottom. The gluteus maximus muscle, judging from its attachments, functioned like man's, as an extensor of the hip. The pelvic ligaments, important in coping with the stresses imposed by bipedalism, were similar to those of man. The lower ends of the femurs show that the knees were close together in walking and standing, that the knee joint could be fully extended—an important aspect of human-type walking—and that the pattern of weight transmission at the knee was manlike and not apelike. The femoral condyles are shaped like man's, and the lateral condyle is larger than the medial. Robinson (personal communication) has estimated that the lower limb was elongated as in man, not short as in the apes. It can therefore be concluded that *Australopithecus* was an erect biped.

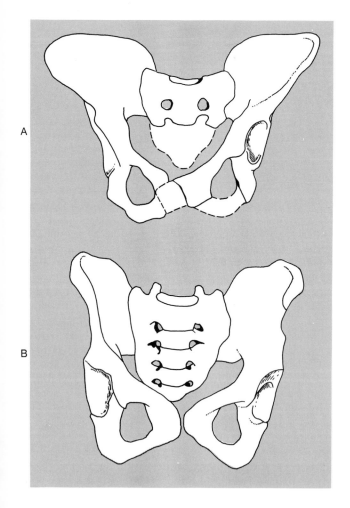

Figure 55 Pelvis of *Australopithecus africanus* (*A*) compared with that of *Homo sapiens* (*B*). (After Robinson, 1963.)

A

B

There are some features of the lower limb that do suggest that bipedalism was not as efficient as in modern man. The small muscles of the hip, the gluteus medius and gluteus minimus, probably did not function with full effectiveness as rotators and stabilizers of the hip. In man these muscles are very important because they maintain the body's center of gravity at a rather constant level in a more or less straight line during walking, thus reducing energy expenditure. *A. africanus* probably walked with toes pointed in, perhaps with some swaying of the trunk, and less efficiently than man. However, this hominid was certainly a biped, and well on the way to being a fully effective one.

The locomotor adaptations of *A. africanus* were unique in both the forelimb and hindlimb. By the end of the Pliocene, this hominid lineage was fully bipedal, although not as well adapted as later men, while still retaining a shoulder girdle that apparently indicateds arm-swinging or suspensory abilities. It is possible that we are dealing here with the later stages of a long and rather slow process of adaptation away from arm-swinging, arboreal, forest life to bipedal, ground-foraging, open-country habits. Whether this process proceeded gradually or in spurts we have at present no way of telling.

Evidence from associated tools, animal remains, and locomotor and dental adaptations suggests that *A. africanus* was probably a bipedal cooperative hunter–gatherer like man and not an individualistic forager like other primates. The presence of tools manufactured to a regular pattern implies that the hominid brain was then capable of "the imposition of arbitrary form," as Holloway (1969) has phrased it, a capacity essential for the development of "culture."

Professor Alan Mann (1968) has made some interesting analyses of tooth wear and maturation in *A. africanus*. He concludes that no individuals lived more than 40 years and that only 15 per cent reached the age of 30 years. The mean age at death was around 20 years, although a third of all individuals lived 10 or more years beyond reproductive age. His analyses of tooth maturation indicate that *A. africanus* matured relatively slowly, like man, and therefore had prolonged periods of infant and juvenile life. This implies that infants would have been dependent on their mothers longer than apes and that the periods of life during which most learning occurs were similarly extended.

KROMDRAAI

The fossil hominids and the mammals come from two separate sites at Kromdraai (Brain, 1958). The mammals indicate an age younger than Swartkrans (Cooke, 1963); from this site have come several stone tools. It has been assumed that the skull site and the faunal site are equivalent in age, although this may well be incorrect.

Only a small number of hominid fossils have been located at Kromdraai, representing no more than five or six individuals. The type specimen of *Paranthropus robustus* consists of parts of the skull and face, and much of the dentition, together with a number of postcranial bones. The teeth (Robinson, 1956) are basically similar to those from Sterkfontein, although larger (particularly broader). The premolars are somewhat more molarized, and the cheek teeth are a little larger relative to the front teeth. Because of this, masticatory muscles were somewhat larger than in Sterkfontein hominids, and the face is more strongly buttressed and shorter than in *A. africanus*. The large teeth and shorter (and probably deeper) and more robust face suggest that this hominid was perhaps a somewhat "better" herbivore than *A. africanus*.

From Kromdraai has come a possibly associated lower humerus and upper ulna, and a talus (Le Gros Clark, 1964). The humerus and ulna closely resemble those of man and differ from those of the apes. No detailed functional analyses of this region of the body have yet been made, so little more can be said about its function. Recently, Patterson and Howells (1967) analyzed the humeral fragment using a multivariate statistical technique, concluding that the specimen came from a form resembling man. The talus, which probably comes from the same individual, was studied by Le Gros Clark (1947). He concluded that the talus was sampled from a biped. According to Le Gros Clark (1964), and to Drs. Michael Day and Bernard Wood (1968), who have recently analyzed this bone using multivariate statistical analysis, the Kromdraai talus differs from those of both man and apes. There are a number of features that suggest it was part of a foot adapted for bipedal weight bearing, but there are also indications that this foot would have had only a poorly developed arch and might well have had a big toe that could be used for grasping. (However, according to Day and Wood, the Kromdraai talus does closely resemble that from early Pleistocene deposits at Olduvai Gorge in East Africa, which was certainly part of a bipedally adapted foot.)

One other postcranial fragment, part of a pelvis, is of interest. It indicates a body size somewhat greater than that of the (supposed) female from Sterkfontein and exhibits a number of features pointing to bipedalism as the main locomotor adaptation.

A third Kromdraai specimen, an immature mandible, is interesting because its first permanent molar (Robinson, 1956) is so small (as small as the smallest from Sterkfontein, and quite a bit smaller than molars from Swartkrans generally assigned to *P. robustus*). This may indicate the presence of a second hominid form at Kromdraai, or merely reflects sexual or size variability.

The Sterkfontein and Kromdraai hominids have been regarded as different genera by Robinson (1961, 1963, 1964). There is some evidence to suggest that they are different species, although the Kromdraai sample is really too small for definite statement. Generic separation is hard to sustain, and accordingly I would follow Tobias

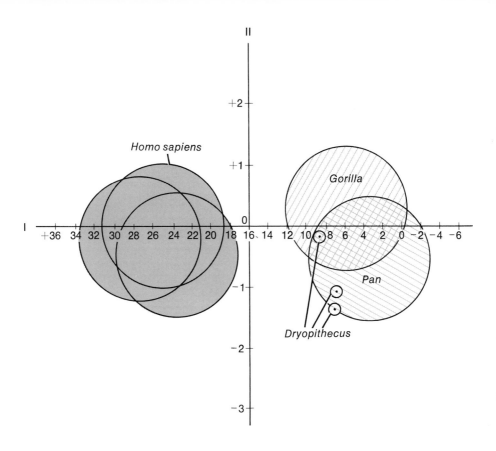

Figure 56 Canonical analysis is a useful multivariate statistical technique (multivariate in that it utilizes more than two variables). In the example here, seven angles and indexes, reflecting so-called functionally significant characters, were taken on the talus (the foot bone that articulates with the tibia) of populations of humans, chimpanzees, and gorillas. These measurements were manipulated by computer, and seven "new" axes were produced. These axes were so designed as to concentrate most of the variability within the original seven-dimensional system into the first new axis, most of the remainder of the variability into the second, and so forth. In this particular example, the first two new (canonical) axes contained almost all of the variability. Thus what was originally seven-dimensional space can be pictured, with minimal distortion, in just two dimensions. The circles encompass populations of bipeds (men) and the knuckle-walking apes. The same seven angles and indexes were obtained from three tali of East African *Dryopithecus* from the early Miocene of Kenya. In terms of the first two canonical variates, these tali are functionally much more similar to those of chimpanzees and gorillas than to those of man. (After Day and Wood, 1969.)

(1967), among others, in treating both samples as species within *Australopithecus*. The Kromdraai material is intermediate in a number of respects between the Sterkfontein and Swartkrans samples. The age of Kromdraai is generally assumed to be younger than Swartkrans, although, as has been noted, this interpretation may be incorrect.

SWARTKRANS

Swartkrans has yielded the largest known sample of early Pleistocene hominids. The site is considered to be younger than Sterkfontein (Cooke, 1963) and is probably 2 million to $2\frac{1}{2}$ million years old. The cave at Swartkrans was originally a large cavern connected to the surface by a long vertical shaft (Brain, 1967a). This opened in the overhang of a cliff, where hominids probably camped. Their remains, together with those of many mammals as well as stone tools, evidently lay on the surface before being washed into the cave. Brain has shown that the proportions of the various bony parts preserved indicate that the animals had been eaten by carnivores—human or otherwise—before their deposition. Two fairly complete yet crushed and distorted skulls have been recovered,

Figure 57 Diagrammatic drawings of the Swartkrans cave at the time of deposition (*A*), and just before excavation (*B*). (After Brain, 1967a.)

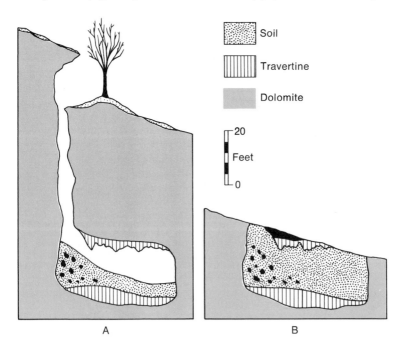

THE ASCENT OF MAN

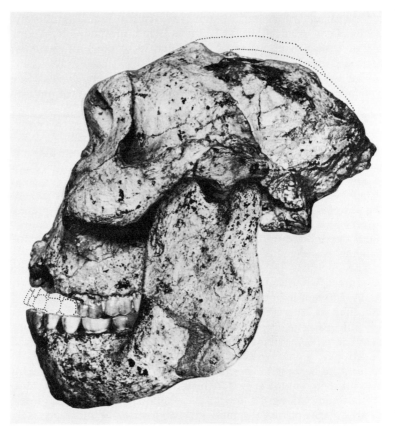

Figure 58 Side view of skull of *Australopithecus robustus* from Swartkrans. Cranium and mandible come from different individuals. (Courtesy of E. L. Simons.)

both said to be from females, and have been reconstructed by Robinson (1954, 1963, 1964). Like most of the Swartkrans sample, these skulls are generally placed in the same species as the Kromdraai material, *A. robustus*. They are robust, powerfully built skulls with well-developed crests and ridges for muscle origins and insertions. Brow ridges and zygomatic arches are bigger than in *A. africanus* and the forehead is less domed. The face is flatter and somewhat deeper. It is impossible to obtain brain volumes from these crushed skulls, although estimates have been published in the past.

A recent discovery, an almost complete endocast that probably belongs to the same species as the skulls, yields a volume of 530 cm^3 (Holloway, 1970), significantly greater than the mean value for *A. africanus*. This may be because of sexual difference (it could be male, and all known *A. africanus* values might be for females) or more probably because of greater body size in *A. robustus*. This endocast is of further interest because it demonstrates that the cerebellum of *A. robustus* was relatively larger than that of apes of equivalent brain size, a significantly human trend (Holloway, personal communication). This probably indicates a greater degree of fine control of movements (hand and locomotor, for example) than in the apes.

The well-sampled dentition of *A. robustus* again shows some differences from *A. africanus*. These can be overemphasized (indeed, have been) and there is overlap between the species, but the basic differences seem to lie in the direction of increased cheek-tooth size in *A. robustus* (Robinson, 1956). The molars are greater in area, and the premolars are more molariform. Selection has evidently favored greater chewing area than in *A. africanus*. The anterior teeth (canines and incisors) have remained as small as in *A. africanus* or may even have become absolutely a little smaller—the size of the samples involved makes it hard to answer this. It is of interest to note that a similar trend—toward increased body and cheek-tooth size and reduced front teeth—characterizes the extinct gelada baboons of subgenus *Theropithecus* (*Simopithecus*).

The contrasts in skull morphology between *A. robustus* and *A. africanus* (Robinson, 1963, 1964) can be related almost exclusively to the differences in cheek-tooth size and masticatory muscles and the changes necessary to increase the mechanical efficiency of chewing. Large teeth require more robust mandibles, which in turn need more powerful masticatory muscles. These must be anchored on stronger bony structures. Thus the zygomatic arch is more robust, and because the brain case is relatively small the attachment area for temporal muscles is increased by the formation of a central crest running along the top of the skull. The face is deeper and flatter so that powerful lateral movements can be more precisely controlled, also because the anterior dentition remains small.

Clearly then, selection has favored larger cheek teeth in *A. robustus* than in *A. africanus*, and from this single change have flowed all the other cranial differences between these two species. The increase in tooth size might be related to increased body size, for there is some evidence to suggest that *A. robustus* was the larger animal, as we shall see when the postcranial remains are discussed later. There is an alternative explanation, Robinson's (1954) dietary hypothesis, which argues that *A. robustus* (or *Paranthropus*, according to him) was a vegetarian feeding on shoots and leaves, berries, tough wild fruits, roots, and bulbs, whereas *A. africanus* had a more nearly omnivorous diet, which may have included a fair proportion of flesh.

However, Tobias (1967) believes that this hypothesis is incorrect. He points out that "the australopithecines . . . differentiated into a series of taxa characterized by differing degrees of enlargement of the cheek teeth and, naturally, of the supporting structure, muscular prominences, masticatory stress columns, and so on. Such differentiation in the size of cheek teeth, of itself, provides no evidence of major ecological or adaptive radiation" (Tobias, 1967:228). This argument may well be wrong, for savanna and gelada baboons, without differences in body size, contrast in tooth size and proportions and skull morphology in much the same ways that *Australopithecus* species do. This also applies to gorillas and chimpanzees, which are of different body size. In both cases, the variations are due to differences in diet: chimpanzees are frugivorous, gorillas herbivorous; geladas feed on small, tough morsels, whereas savanna baboons have a more varied (mainly) vegetable diet.

A number of questions need to be asked at this point. First, are the differences between *A. africanus* and *A. robustus* sufficient to indicate different feeding habits? Certainly, the gelada example suggests an explanation for the differences; that is, *A. robustus* was more of a herbivore, feeding on more small, tough morsels. However, different populations of the same species (it may even be subspecies) of the baboon *Papio cynocephalus* differ more in tooth area and skull proportions than do the two *Australopithecus* species (Freedman, 1963). These differences are apparently due to body size contrasts, and it is not known what factors affect this. So *A. robustus* and *A. africanus* could have had slightly different diets or their differences could be due to body size. This question can be left open for the present.

Second, could *A. africanus* have been ancestral to *A. robustus*? Robinson (1954) believes that this is impossible and that they represent two quite distinctive adaptive types that had been separate for a considerable period of time. However, the differences between *A. africanus* and *A. robustus* seem to have been overstressed; many of them can be matched in living species. And the differences, at least in the skull, appear to be part of one functionally interrelated complex. Also, *A. africanus* comes from deposits that are all older than the *A. robustus* sites (Cooke, 1963), so on stratigraphic grounds there are no objections to an ancestor–descendant relationship. Finally, there are few characteristics that would indicate that *A. robustus* was not descended from *A. africanus*. That is, it is relatively easy to explain the differences between them in functional, morphogenetic, and selective terms. It seems at least a possibility then that they are sampled from different portions of a single lineage. If so, a further question can be asked.

Should they be classified into different species? Once again the variability between the two groups can be matched both within living species, like savanna baboons, and between species, for example, savanna baboons and geladas. For the present, it is probably most appropriate to retain them in separate species; recent evidence suggests they came from separate lineages.

A number of postcranial bones have come from Swartkrans, and these all probably belong to *A. robustus*. Part of a rather crushed and distorted pelvis shows that at least some of the Swartkrans hominids were larger than the *A. africanus* individuals for which pelvises are known (Napier, 1964). (However, it should be emphasized that the associated vertebral column and pelvis from Sterkfontein on which much of the body weight estimates for *A. africanus* have been based probably belonged to a female.) There appear to be some morphological contrasts with the Sterkfontein material that

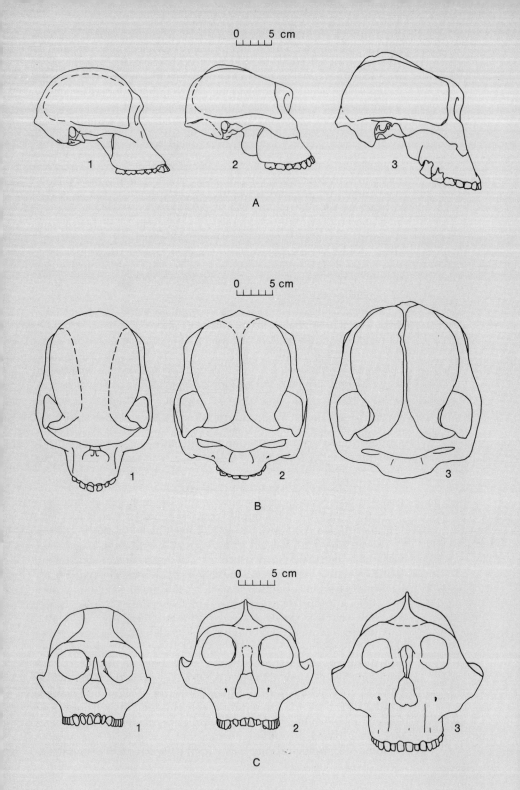

A

B

C

Figure 59 (opposite) Lateral (*A*), superior (*B*), and frontal (*C*) views of *Australopithecus africanus* (*1*), *A. robustus* (*2*), and *A. boisei* (*3*). Of particular interest are the cranial characteristics that change with the increasing cheek-tooth size (*A. africanus* having the smallest teeth, *A. boisei* the largest). Considered as a series, the face is flattest and deepest in *A. boisei*, crests and bony buttresses are better developed in *A. boisei*, and the temporal fossa (indicating temporal muscle bulk) is largest in *A. boisei*. All these features appear to be related to the size and mechanical efficiency of the posterior dentition. (After Tobias, 1967.)

are probably not due to crushing during fossilization. Thus the ischium is relatively longer in *A. robustus* than in *A. africanus*, and the acetabulum and femoral head are relatively smaller (Zihlman, 1967).

Two proximal femurs are known and show a number of contrasts with the femur of modern man; the head is smaller and the neck longer and more buttressed below. Dr. John Napier (1964) has argued that this implies a less perfect bipedalism in *A. robustus* than in *A. africanus* or man. Yet because of the sampling involved it is very difficult to compare the two *Australopithecus* species; in comparable parts they appear basically similar, except for a few pelvic differences that are poorly understood. The Swartkrans material reinforces the evidence from Sterkfontein in indicating that the *Australopithecus* species were somewhat less well-adapted bipeds than man. In both species, acetabula face laterally rather than laterally and forward; femoral necks are long and buttressed, and

Figure 60 Pelvis of *Australopithecus robustus* from Swartkrans. (Courtesy of J. T. Robinson.)

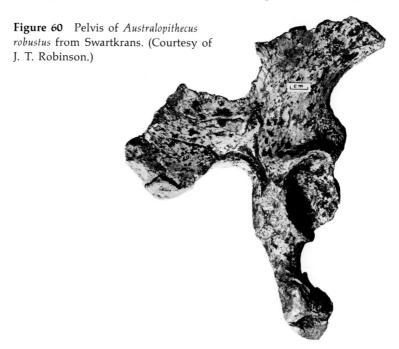

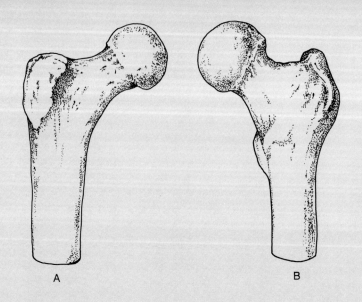

A   B

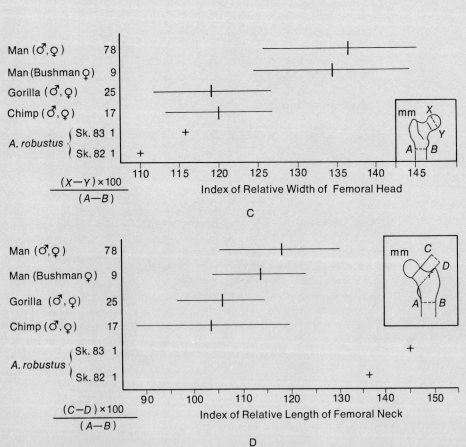

$$\frac{(X-Y) \times 100}{(A-B)}$$

Index of Relative Width of Femoral Head

C

$$\frac{(C-D) \times 100}{(A-B)}$$

Index of Relative Length of Femoral Neck

D

Figure 61 (opposite) Femoral neck and head of *Australopithecus robustus* (A) and *Homo sapiens* (B). Differences that are particularly well marked are the relative size of the femoral head and the relative length of the neck. These two features are probably interrelated and, together with other characters, suggest that *A. robustus* was a less well-adapted biped than *H. sapiens*. The graphs (C and D) show means and standard deviation ranges. (After Napier, 1964.)

heads are relatively small. Explanations for some of these features have been produced. For example, the femoral head may be small because *Australopithecus* was still in the process of adapting to habitual erect posture (this view is supported by the relatively small sacrum and sacroiliac articulation). The neck might be long, first, because the head is small and, second, to increase the breadth between the proximal ends of the femurs (because the pelvis was narrow), thus increasing stride length (Walker, personal communication). Until more work can be done it is probably not worthwhile discussing this matter at length. The unsatisfactory postcranial evidence is the major reason for not being more definite about the relationships of *A. africanus* and *A. robustus*.

Approximate body weight for the Swartkrans hominid can be estimated, based on pelvic size and femoral shaft diameter. *A. robustus* was probably on average a larger animal than *A. africanus*. Whereas the extreme weight limits for *A. africanus* were probably around 40 to 70 pounds, *A. robustus* may have weighed between 80 and 140 pounds.

Two hand bones have come from Swartkrans, a broken fourth metacarpal and complete first metacarpal (Napier, 1959). Both are smaller than average human bones and do not resemble those of apes. They evidently come from a human-type hand that was nevertheless capable of powerful gripping—one that retains features suggesting a previously arboreal but not, according to Tuttle and Oxnard, knuckle-walking history. Napier believes that the hand bones come from different hominid species, one *A. robustus* and the other more "advanced," because the combination of morphological features seen in the two bones cannot be matched in modern hands. However, this could be because the hand of *A. robustus* was not like that of any known primate, a not unsurprising conclusion from what we have seen from the rest of the skeleton. This is a good example of so-called mosaic evolution, the concept that different parts of a single functional complex need not evolve at the same rate.

It has been claimed that a second hominid is represented at Swartkrans, one more closely allied to *A. africanus* or to *Homo*. This was originally referred to as *Telanthropus capensis* (Broom and Robinson, 1949), although all workers now agree that this does not correctly describe the material. As originally constituted, *"Telanthropus"* consisted of a couple of mandibles and a piece of maxilla. Recently, Ronald Clarke, F. Clark Howell, and C. K. Brain (Clarke,

Howell, and Brain, 1970) have been able to articulate the maxilla with portions of a cranial vault previously regarded as *A. robustus.* The resulting specimen differs somewhat from the other Swartkrans hominids. The teeth are smaller and the jaws and face less robust. The cranium, as far as it is preserved, is more gracile and rounded, with less robust crests and ridges. There is a distinct possibility that this represents a second hominid species at Swartkrans. The dentition resembles that of *Homo erectus* in size, although the brain was apparently small and the skull vault bones thin, both non-*erectus* characteristics. This could represent a descendant of *A. africanus* (equivalent to *"Homo habilis"* in East Africa) and might be classified with that species. Alternatively, it might be better to wait for more material before assigning it to a species.

An alternative explanation is that *"Telanthropus"* is simply a small and gracile *A. robustus* specimen at the extreme lower end of the range of variation (Wolpoff, 1968), Several new specimens from Swartkrans (Brain, 1970) to some extent fill in the size gap between *"Telanthropus"* and *A. robustus,* and the Kromdraai juvenile mandible does have a very small first molar, smaller than that of *"Telanthropus."* This matter requires more detailed statistical study, particularly multivariate analysis, before it can be satisfactorily settled, although it does now appear probable that two species are represented at Swartkrans.

MAKAPANSGAT

The hominid sample from Makapansgat is of moderate size, smaller than for Sterkfontein, but nevertheless adequate. An extensive fauna is known, and may indicate an age somewhat older than Sterkfontein. The first specimen recovered, in 1947, was a piece of skull comprising the occipital bone and parts of the parietals (Dart, 1948). Later, a more complete skull was found; this is said to be female and resembles the skulls from Sterkfontein quite closely (Dart, 1962). For this and other reasons the Makapansgat material has been generally assigned to *A. africanus.* The markings for the temporal muscle are quite distinct on the first specimen; if this is a male, it is quite likely that males of this species would have had central cranial crests like *A. robustus* did. The brain volume of the more complete specimen has been estimated at 435 cm^3 (Holloway, 1970). If this is added to the four volumes from Sterkfontein and one from Taung, then the mean brain volume for *A. africanus* comes to around 440 cm^3 (Holloway, 1970). It is relatively easy statistically to calculate the probable range of volumes for the population from which these skulls were sampled. It is very unlikely that any individuals in the population would have had volumes greater than around 575 cm^3 or less than 300 cm^3. However, this South African population would probably not constitute the entire

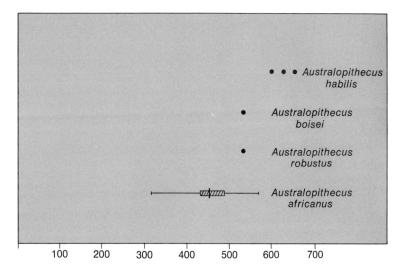

Figure 62 Brain volumes of *Australopithecus* species. In the case of *A. africanus,* the vertical line represents the sample mean, the hatched area the known range, and the horizontal line the estimated population range. Bottom numbers are brain volumes in cm³. (Pilbeam.)

species, because we know that terrestrial primates are generally very wide-ranging animals. The South African hominids are probably sampled from a subspecies lineage of a "species" much more widely distributed throughout Africa, and perhaps Asia, too.

Some stone objects have been located at Makapansgat that may be tools; mostly they seem to have been used for crushing and pounding, probably in the preparation of vegetable foods or the comminution of bone to obtain marrow. Dart has proposed that many of the animal bones and jaws associated with *A. africanus* at Makapansgat were in fact utilized as tools: dentitions as saws and scrapers, long bones as clubs, and so forth. Dart (1957) originally proposed this idea, that bones and teeth had been used as tools ("osteodontokeratic" tools), because of the fact that certain animal parts turned up regularly whereas others were extremely uncommon. For example, lower jaws were very common, tail vertebrae were almost nonexistent, and forelimb bones were five times more frequent than hindlimbs. How could these distributions be explained? Dart believed that the differential preservation reflected differential use of certain body parts as tools. Recent experimental work by Brain (1967b, 1970) has thrown considerable doubt on this whole issue. Brain scattered animal carcasses in Hottentot villages, where the remains were eaten by dogs, and discovered that the effects of scavenging and the differential weathering of delicate bones such as the shoulder blade could account for the distributions that Dart had obtained for the Makapansgat material.

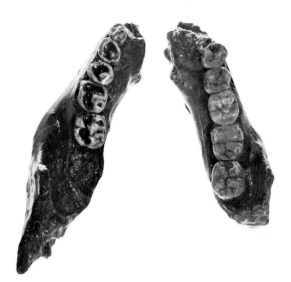

Figure 63 Casts of lower jaws of *Australopithecus africanus* (left) from Makapans and the type specimen of *A. robustus* from Kromdraai (right) showing the similarities between the two. (Pilbeam.)

The collections at Makapansgat may well have been made mostly by hominid hunters but the remains once discarded were probably chewed by other carnivores afterward.

Most of the animal remains were antelope; evidently, *A. africanus* had a penchant for venison. Of these, many were medium to large forms, so evidently *A. africanus* was an effective hunter despite small brain and body size.

As already noted, the skulls from Makapansgat are essentially similar to those from Sterkfontein and apparently do not indicate the more extreme herbivore features of *A. robustus.* Jaws and teeth are well known, and these also follow basically the pattern seen at Sterkfontein, although there are some differences between the samples that may reflect population differences (or may simply be because of the relatively poor samples). Jaws are robust and well buttressed, teeth large and broad. In some mandibular and dental dimensions, the Makapansgat sample falls between Sterkfontein and Swartkrans, though this could be due to sampling error of course. However, in a fair number of important (or so it is thought) features, Makapansgat is close to Sterkfontein. For example, premolar morphology and jaw shape are both typical of *A. africanus.*

At least three alternative explanations can be offered for this situation. First, Makapansgat may sample a commingling of *A. robustus* and *A. africanus* lineages. This assumes that these lineages were sympatric (more or less) and synchronic, also that they were probably not separate species (although species of baboons—*Papio cynocephalus* and *Papio hamadryas*—do form hybrids in their narrow contact zone). Tobias (1967) has tentatively proposed this hypothesis. Second, Makapansgat could be an intermediate population from the same lineage as both *A. africanus* and *A. robustus,* but closer in time and morphology to the former and therefore more similar to the Sterkfontein hominids. This hypothesis I believe to be more likely than the first. Third, assuming that *A. africanus* and *A. robustus* were separate lineages, Makapansgat is sampled from the *A. africanus* lineage but the specimens are larger-toothed for reasons that are at present poorly understood (ecological differences in habitat, for example). As an alternative, this would rank ahead of the second in terms of probability.

SUMMARY OF THE SOUTH AFRICAN HOMINIDS

Five South African sites have yielded early hominids. Almost certainly, the samples are drawn from populations of species that were much more widely ranging throughout Africa. The sites span a considerable period of time, perhaps as much as 1 million or more years. None of them appear to have been actual living sites, although hominids probably did camp close to cave entrances. The local habitats during this time period were almost certainly grassland and open woodland; associated animal remains, together with stone and occasional bone tools, suggest that these early human forerunners were open-country hunters as well as gatherers of vegetable material.

There are two alternative evolutionary explanations for this material; either two lineages—*A. africanus* and *A. robustus*—are involved, or the former species evolved into the latter. *A. robustus* was probably a larger form than *A. africanus* and may well have been feeding on many small items of vegetable food. Although both species had small brains, the brain of *A. robustus* was probably significantly larger than that of *A. africanus,* a feature perhaps related to its greater body size. There is some evidence to suggest that the brain was organized in a distinctly nonpongid way, resembling that of hominids, and this fact, together with the evidence for tool-making behavior, implies that these early hominids were becoming or had become "cultural" animals; they were more like men than like apes. Recent material from East Africa supports the view that two lineages are involved; however, it is possible that both lineages have been sampled at sites other than Swartkrans.

6 East African Early Hominids

Although a few fragments of hominids of reputed "early Pleistocene" age were known from East Africa before 1959, since that time many more specimens, in most cases well dated, have been recovered from sites in Kenya, Tanzania, and Ethiopia. Of principal interest are the magnificent series of discoveries made since 1959 by Drs. L. S. B. and M. D. Leakey at Olduvai Gorge and those made more recently by Professor F. Clark Howell in the Omo region of Ethiopia and by Richard Leakey in the East Rudolf area of Northern Kenya.

The remains fall into two major and quite distinct hominid lineages, one gracile, the other robust, both apparently with considerable time duration. In 1959, the Leakeys reported a new type of robust hominid, similar in some ways to *A. robustus*, which they named *Zinjanthropus boisei* (Leakey, 1959). Five years later, Leakey, together with Napier and Tobias, described a second East African hominid, similar to *A. africanus* but more advanced in certain features (Leakey, Tobias, and Napier, 1964). This they called *Homo habilis*. In 1967, Tobias described *Z. boisei* in detail, transferring it to *Australopithecus*. This classification will be followed here. There has been considerable debate concerning the status of *H. habilis* since its initial description; I believe that the present evidence favors its description as a species of *Australopithecus*, *A. habilis* (Simons, Pilbeam and Ettel, 1969). In the following account, this taxonomic scheme will be followed.

OLDUVAI

Olduvai Gorge is a miniature Grand Canyon, cut into the Serengeti Plain of Tanzania (Leakey, 1965). Beginning early in the Pleistocene, volcanoes of Ngorongoro and Lemagrut erupted a series of basalts, volcanic tuffs, and ashes. A lake was present in the Olduvai region

Figure 64 The relative time placement for hominid-bearing sites in South and East Africa. (Pilbeam.)

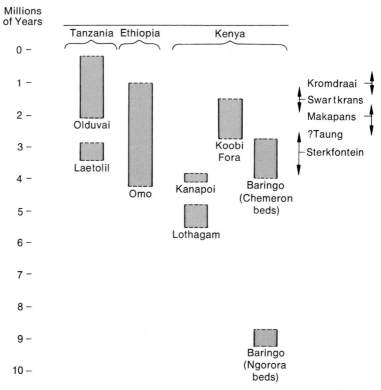

Note added in the final stage of proof: The South African sites are now thought to range in age from 2 million to 3 million years (Cooke, personal communication); Makapans is probably the oldest, Kromdraai or Swartkrans the youngest.

for much of the time period represented by exposed rocks in the Gorge. Late in the Pleistocene, river action cut through the accumulated deposits down to the earliest basalts, exposing a geological sequence that runs almost continuously from the present back to nearly 2 million years ago. The lake apparently provided an attractive focus for all kinds of mammals including hominids. As far as can be reconstructed, the local habitat was open woodland and parkland, probably with grass close to the lakeside. A number of hominid living sites have been discovered close to what was the lake's edge, and the evidence indicates that these were home bases—possibly seasonal camps—from which the hominids hunted.

The Olduvai Gorge sequence was first studied by Leakey in 1931. In 1951, Reck divided up the deposits into a series of beds, Bed I being the oldest. Recently, geologists at the University of California at Berkeley and elsewhere have obtained a series of radiometric dates for the lower parts of the Bed I sequence (Evernden and Curtis, 1965). Bed I is mostly volcanic in origin (Hay, 1967), although there are stream and lake deposits, too. The earliest part of Bed I, a basalt flow, has been dated at around 2 million years. The oldest living floors at Olduvai closely overlie the basalt. Some way up Bed I, overlying several important hominid sites, is a datable horizon with an age of around $1\frac{3}{4}$ million years. About half way up Bed I is so-called Marker Bed A, also with a date of some $1\frac{3}{4}$ million years. The top of Bed I is traditionally defined by Marker Bed B, exposed in various parts of the Gorge; this is not dated. Bed II conformably follows Bed I and consists mainly of waterlain

Figure 65 Diagrammatic representation of Olduvai Gorge Beds I and II showing the approximate placement of hominids. (After M. D. Leakey, 1967.)

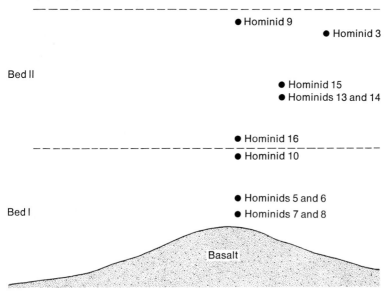

stream and lake deposits. In many places it is as thick as Bed I, but it took very much longer to accumulate, because the deposits are mainly lacustrine rather than volcanic. This cannot yet be certainly decided, because no reliable dates are available above Marker Bed A until the very top of the sequence (Bed V). An age of $\frac{1}{2}$ million years for the top of Bed II is now considered to be unreliable, and the real age may be close to 1 million years (Isaac, 1969).

The lower parts of Bed II are faunally similar to Bed I, although upper Bed I and lower Bed II contain some rather more evolved elements than lower Bed I. The upper part of Bed II has yielded a rather different fauna that is said to be typically "early middle Pleistocene," although such a term has probably only local significance and therefore rather restricted usefulness (Leakey, 1965). Lower Bed II and Bed I faunas are "early Pleistocene." The age of the faunal transition from "early" to "middle" Pleistocene in East Africa is unknown, although there is some indication from deposits at Lake Natron in Tanzania that sediments of upper Bed II almost certainly fall in age between $1\frac{1}{2}$ million and 1 million years (Isaac, 1969). Beds III and IV are probably also "middle Pleistocene," although obviously younger than Bed II.

The stone tools of Bed I and lower Bed II are of Oldowan type. This industry was once colloquially described as a "pebble tool industry" after the characteristic small chopper-type tools. However, recent work by Dr. Mary Leakey (1967) has shown that the Oldowan is a much more complex assemblage than was previously thought. Although it does contain choppers and pounding tools—probably utilized in digging, crushing, and pounding vegetable food and bone—other small tools are also found, for example, scraping and cutting tools that were probably part of meat-preparing tool kits. These tools have been collected from living floors that became sealed in by lava flows and ash falls, and thus a good sampling of both tools and animals taken for food has been preserved. (This was not the case for South Africa, where no actual living sites have been discovered.)

According to Mary Leakey (1967), in upper Bed II the Oldowan culture persisted, in somewhat evolved form, alongside industries with hand axes, tools made on blanks more than 10 cm long. She believes that these represent two distinct cultural traditions possibly made by two different hominid species and that the Acheulean (hand-axe) cultures were brought into the Olduvai region from elsewhere. Dr. Glynn Isaac (1969) has suggested an alternative possibility, that the sudden appearance of hand axes might be due to the invention and spread, in a very short time, of techniques that permitted the production of flakes large enough for this kind of tool. The presence or absence of axes in various assemblages might be due to local variation or to differing functions of tool kits.

According to L. S. B. Leakey (1966) and some other workers, there are three lineages of hominids represented at Olduvai.

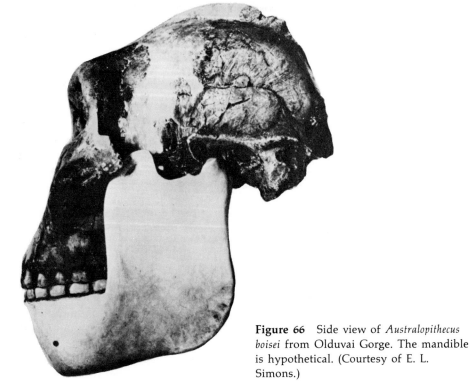

Figure 66 Side view of *Australopithecus boisei* from Olduvai Gorge. The mandible is hypothetical. (Courtesy of E. L. Simons.)

I believe it probable that only two are present, and we shall be discussing the hominids within the latter framework, although mentioning alternative hypotheses at the appropriate time.

The first lineage is a robust species of *Australopithecus, A. boisei,* represented by two milk teeth from a site of late Bed II age (Hominid 3), some upper teeth from middle Bed II (Hominid 15), an almost complete cranium and upper dentition from lower Bed I (Hominid 5), and a femoral fragment from upper Bed I or lower Bed II (Hominid 20). Recently, Richard Leakey has recovered skulls of this species from the East Rudolf area of Northern Kenya. The milk teeth were found in 1954, the skull and femur in 1959. The cranium (Hominid 5) was originally made the type of a new genus and species, *Zinjanthropus boisei,* now known generally as *A. boisei.* It is one of the most complete *Australopithecus* skulls found, being almost undistorted, and has been exhaustively described by Tobias (1967). It is extremely massive and robust, with very large teeth. When it was first described, a number of workers considered it to be close to *A. robustus,* but in certain features—for example, cheek-tooth area—*A. africanus* and *A. robustus* are more similar than *A. robustus* and *A. boisei.* Basically, the distinctive cranial features of *A. boisei* are related to the size and mechanical efficiency of the cheek teeth. It may well be a descendant of *A. robustus.*

Molars and premolars are enormous relative to incisors and

canines and are noticeably broad compared with those of other *Australopithecus* species. Selection pressures were clearly favoring increased tooth chewing area, yet mechanical demands prevented the dentition as a whole from being too long from front to back, hence the lateral broadening. The teeth are set in a deep, short, and massively buttressed face, clearly designed to cope with considerable chewing stresses. The brow ridges are very prominent. The face is short probably to reduce the load arm of the incisors, thus allowing more power to be applied by the slicing incisors and canines. Facial depth is correlated with the height of the ascending ramus of the mandible. All herbivores have high rami, to increase the moment arm of the masseter and pterygoid muscles (the main muscles of mastication in herbivores) around the jaw articulation and also to provide a large area for the insertion of these important muscles. *A. boisei* was clearly a very efficient herbivore. The moment arm of the masseter was further increased in this form because its origin had shifted forward to the very front of the zygomatic arch, which itself is relatively far forward along the tooth row (above the last premolars) as the dentition and face have been "tucked in" under the brain case. The area of origin for the masseter on the zygomatic arches is deeply excavated, indicating the powerful nature of this muscle.

Figure 67 The approximate force vectors of the masseters and temporals in *Homo sapiens* (A) and *Australopithecus boisei* (B). Note the long lever arms of both muscles, an adaptation for power rather than speed. The line *ab* is the "masticatory axis" and connects the jaw joint with the midpoint of the functional cheek-tooth area. The force vector of the masseter intersects this more than halfway along *ab*, indicating that chewing force will be greater (considerably greater in *A. boisei*) than the force exerted at the jaw joint. This is once again an adaptation for powerful chewing. (After Crompton and Hiiemäe, 1969.)

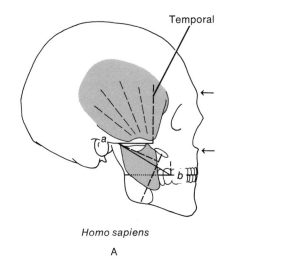

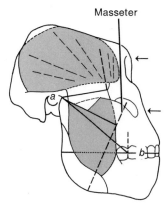

Temporal

Masseter

Homo sapiens

A

Australopithecus boisei

B

The temporal muscles were also well developed, as is evidenced by the presence of a sagittal crest. The presence of the crest is also influenced by brain size for had the brain and brain case been more extensive no crest would have been necessary to provide extra attachment areas. The anterior parts of the temporal muscle were strongly developed, as is evidenced by the buttressing and cresting behind the brow ridges and on the frontal bone. In gorillas the sagittal crests are most prominent further back, indicating stronger development of the posterior parts of the temporal muscle. Because of this strong anterior concentration in *A. boisei*, the force vector (main line of pull) of the temporal would have been much more vertical than in other primates and more nearly parallel to the vector of the masseter. The resultant of the masseter and temporal vectors was approximately vertical to the tooth row. Thus both sets of muscles were in a mechanically advantageous situation to move the mandible powerfully in almost any direction under precise control. Chewing in hominids, with their flat-crowned teeth, thus can involve movements in all directions, not just fore and aft or side to side as in other herbivores.

The massive temporals "caused" the brain case to be moved back from the face, constricted the frontal region, and increased the size of the temporal fossa. An association of massive brow ridges and backwardly shifted brain case results in the almost total absence of forehead in *A. boisei*.

Thus although *A. boisei* appears to differ in many ways from *A. africanus* and *A. robustus* the differences can all be related to a single functional feature—increase in cheek-tooth size. This was presumably brought about by selective pressures favoring a change to feeding basically on numerous small, tough vegetable items. Jolly's gelada model (1970) therefore helps to explain tooth proportions (small front teeth and large cheek teeth), cheek-tooth size, and all the other correlated features. As Tobias (1967) has shown, the brain cases of *Australopithecus* species are essentially "variations on a theme," and this can be extended to other parts of the skull and dentition. Because these *Australopithecus* species are "variations on a theme," it seems best to classify then into a single genus. In short, with changing selection pressures along a gradient of vegetarian diets from "omnivorous–frugivorous" to "small-item herbivorous" we can probably account for all the cranial and dental differences among *A. africanus, A. robustus,* and *A. boisei,* because most if not all of the changes appear to be related to differences in cheek-tooth size. Thus the different species of *Australopithecus* probably occupied different open-country ecological niches. However, unlike the South African situation, in East Africa two hominid species were living side by side in the same habitat at the same time.

The *A. boisei* cranium is reliably dated to be around 1.8 million years old. The same lineage is apparently represented some hundred thousand or so years later by a proximal fragment of femur from the top of Bed I (Day, 1969a) or more probably the lower parts of Bed II. This specimen preserves the upper part of the shaft, the neck, and most of both trochanters; it is very similar in size and form to the two proximal femurs from Swartkrans. This would suggest the *A. boisei* and *A. robustus* were of approximately similar body size, further supporting evidence that their differences in dental and skull morphology are due to factors other than similar stature, and that they belong to one lineage.

The femoral fragment (Hominid 20) has been described by Dr. Michael Day (1969a). He concludes that *A. boisei* was a habitually erect biped but less well-adapted to bipedalism than *H. erectus* or *H. sapiens*. The femoral neck is long and well buttressed, the hip joint faces laterally, and the areas of attachment for the small glutei are restricted, indicating that the alternating pelvic support mechanism was less evolved than in *H. sapiens*. Day concludes that there are indications from the femur that the center of gravity of the body lay anterior to the hip joint in *A. boisei*. In man it is behind the hip joint, and the posterior torque of the body tends to over-extend the hip; for this reason, the iliofemoral ligament, connecting the pelvis and femur anteriorly, is thickened and leaves prominent attachment markings on both bones. Such markings are absent, or at best poorly developed, in both *A. boisei* and *A. robustus*. Were the center of gravity anterior to the hip joint in these two species, Day believes that it would have been advantageous to elongate the ischial segment of the pelvis, thus increasing the mechanical advantage of the hip extensors attached to the ischium.

The deciduous teeth from late in Bed II times that probably belong to *A. boisei* indicate that this species was present in the Olduvai region for upward of $\frac{1}{2}$ million years (Leakey, 1958). As we shall see, the same lineage lived in East Africa more than 1 million years before its first occurrence in the Gorge.

The brain volume of the single *A. boisei* skull was 530 cm^3 (Tobias, 1967), statistically significantly greater than the older and smaller form *A. africanus*, and the same size as the single estimate for *A. robustus*. The brain cast of *A. boisei* has recently been examined by both Tobias (1967) and Holloway (personal communication). They believe that there is evidence for parietal lobe expansion as well as for cerebellar reorganization. Although *A. boisei* has generally not been regarded as a tool maker, there is no reason to assume that it was not; indeed, if anything the brain evidence supports the view that it was capable of fabricating implements.

In 1964, Leakey, Tobias, and Napier described a second hominid from Olduvai, which they called *"Homo"* habilis (type specimen, Olduvai Hominid 7). This will be described here as *A. habilis*, for reasons to be explained later. The type specimen of *A. habilis* comes from a site stratigraphically somewhat lower than that of the *A. boisei* type, although they can be regarded as essentially contemporaries. Specimens referable to the *A. habilis* lineage have been found throughout Bed I and into the middle of Bed II, spanning more or less the same time period as *A. boisei*.

The type of *A. habilis* consists of a juvenile mandible with teeth, an upper molar, parts of both parietals, and some hand bones. The

Figure 68 Type mandible of *Australopithecus habilis* (right) from Olduvai Gorge. The hominid on the left comes from early Pleistocene deposits in Java. (Courtesy of P. V. Tobias.)

mandible and dentition are most similar to those of *A. africanus* from Sterkfontein. However, the cheek teeth, the premolars in particular, are somewhat narrower than those of the South African hominid, and this feature is confirmed by other teeth of *A. habilis* from the Gorge (Tobias, 1966). The differences are hardly greater than those to be found between subspecies, or at most between two closely related species.

The parietals come from a skull that was either larger or differently proportioned than the skull of *A. africanus*. Although brain volume estimates based on such fragmentary material are bound to be relatively inaccurate, it can be stated that there is a good probability that the volume was greater than 650 cm^3 (Tobias, 1968) and that this value is highly significantly different statistically from *A. africanus* volumes. It is possible that the brain of *A. habilis* was further expanded in the parietal region than the brain of *A. africanus*. Although comparable parts are almost nonexistent, it seems clear that *A. africanus* and *A. habilis* were similar in body size and probable that they were quite alike postcranially.

The hand bones assigned to the *A. habilis* type together with those from an adult individual of the same species from the same site (Hominid 8) have been studied by Napier (1962b). He concludes that the hand was basically manlike with a broad spatulate thumb terminal phalanx. Nevertheless, there were some differences from the human hand. The phalanges were more curved and the thumb was possibly a little shorter than in modern man. Evidently, *A. habilis* was capable of a very powerful grip, for there are well-developed fibrotendinous markings on the bones. Whether or not the fine movements typical of man were also possible is difficult to say from the hand bones alone because delicate hand functions are probably determined as much by the organization of the brain as by the structure of the hand.

This same adult *A. habilis* is also represented by most of a foot (Day and Napier, 1964) and by the clavicle (Oxnard, 1969a, b). The foot is clearly that of a small biped, perhaps no more than 4 to $4\frac{1}{2}$ feet tall. There was a transverse arch as well as the typical primate longitudinal arch, as is evidenced by the shape of the bones and by markings and grooves for the ligaments and muscles that provide support for the arch system. The first metatarsal was parallel to the others, articulating at its base with the second metatarsal. The lever proportions of the foot are more similar to those of man than to those of nonbipedal primates. Power was applied and weight lifted as in man. The first and fifth metatarsals are robust, indicating a human type of weight bearing and transmission, and the talus and calcaneum are articulated and oriented like man's, implying similar weight transmission.

There are some features in which the foot does not resemble man's. For example, the third metatarsal is unusually robust and the axis of rotation of the ankle joint was practically stationary as in only 3 per cent of human feet. When the foot was in resting

Figure 69   Foot bones of *Australopithecus habilis* from Olduvai Gorge. (Courtesy of P. V. Tobias.)

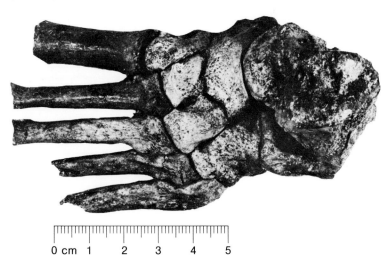

0 cm 1   2   3   4   5

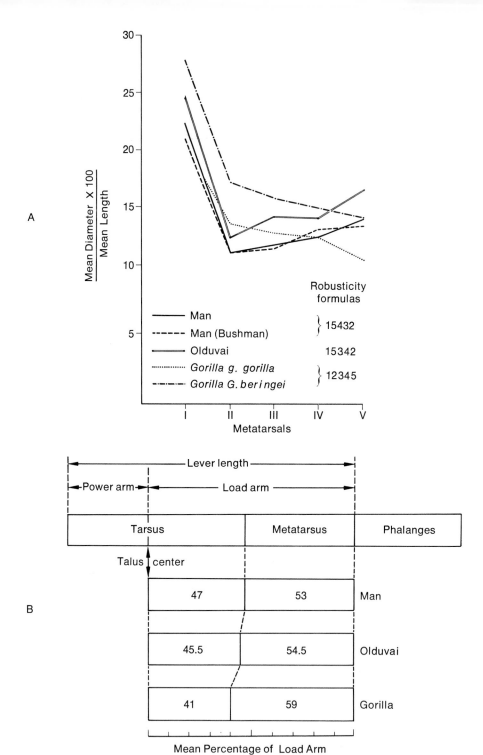

138

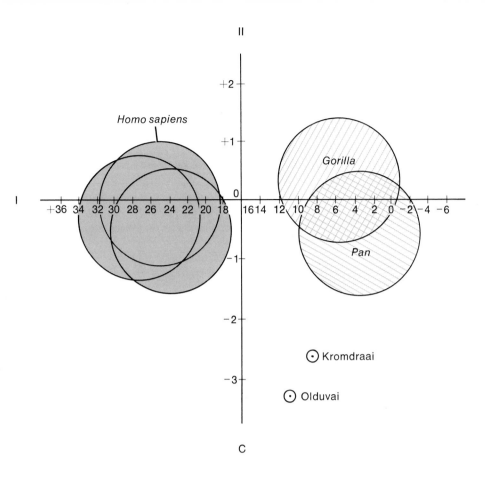

C

Figure 70 *A*: Metatarsal robusticity diagram for the Olduvai foot compared with those of *Homo sapiens* and *Gorilla gorilla*. The Olduvai foot has a relatively more robust third metatarsal than that of most modern men. *B*: The relative contributions of tarsus and metatarsus to the load arm in man, the Olduvai foot, and gorilla. The Olduvai foot may be primitive in this feature, although it is closely similar to that of man. *C*: Canonical analysis of seven talar indexes and angles of supposed functional significance. Hominid tali from Kromdraai and Olduvai are included and are clearly functionally distinct from those of both bipeds and knuckle-walking quadrupeds. However, it is known that the Olduvai talus at least comes from a bipedal foot. (*A* and *B*: after Day and Napier, 1964. *C*: After Day and Wood, 1968.)

stance it was apparently somewhat flexed, suggesting a rather bent-kneed posture. Finally, detailed analysis of the talus shows that it differs from that of modern man in a number of ways. Day and Wood (1968) have analyzed the tali of men and the African apes by canonical analysis, a multivariate statistical technique. They selected seven angles and indexes that demonstrated functionally important ways in which the tali of apes and men differed, then combined these into new functions using a computer. They were able to show a clear distinction between bipedal striding men and pronograde quadrupedal apes. The Olduvai talus, together with the Kromdraai talus that they analyzed at the same time, resembled neither group. Both were closer to men, and they also were close to each other, being no further apart than human subspecies.

It is known that the Olduvai talus is part of a bipedal foot, yet a number of characteristics, particularly those of the talus, indicate that *A. habilis* was not a biped like man. As we have seen, this was also the case with *A. africanus*

The clavicle from (or thought to be from) the same hominid (Hominid 8) has been described by Napier and by Oxnard. Napier concluded that the clavicle was essentially similar to modern man's, although there were a few differences. Particularly noticeable was the increased torsion of the bone, implying that the scapulae were set relatively higher than in modern man and were more like the condition in the apes. Oxnard (1969a) has concluded that the Olduvai clavicle and the Sterkfontein scapula come from forms that were essentially similar in their shoulder girdle morphology and that both species (*A. africanus* and *A. habilis*) or their ancestors were capable of arm raising and suspension. Once again it appears that there are strong postcranial resemblances between *A. africanus* and *A. habilis,* in ways in which both differ from man.

The clavicle and foot of Hominid 8, if they are indeed associated, can tell us something about body proportions in *A. habilis*. Relative to foot size, the clavicle of *A. habilis* is some 10 to 20 per cent larger than is the case for man. Thus it is likely that the forelimb and chest were relatively a little more robust in *Australopithecus* than in modern man, a conclusion drawn already from the Sterkfontein material.

Olduvai Hominid 6 is from the same site as the *A. boisei* skull, consisting of skull and tooth fragments with possibly associated parts of tibia and fibula (Davis, 1964). It seems unlikely on grounds of size that the leg bones belong to the robust hominid, and they are accordingly treated as *A. habilis*. Once again they confirm the fact that this was a bipedal form. The ankle joint is similar in almost every detail to that of man. However, the knee joint shows a number of differences. These may be the result of different patterns of movements; possibly the joint was still in the process of evolving the locking in extension characteristic of man. However, the lower ends of femurs from Sterkfontein indicate that full knee extension was possible in *A. africanus*. Perhaps this is an example of morphology lagging behind behavior, so to speak, and a further instance

THE ASCENT OF MAN

Figure 71 Outline drawing of the phalange (Hominid 10) from Olduvai Gorge (*A*), along with a canonical analysis (*B*) based on nine functionally significant angles and indexes. Hominid 10 is clearly from a bipedal form. (*A*: After Day and Napier, 1964. *B*: After Day, 1967.)

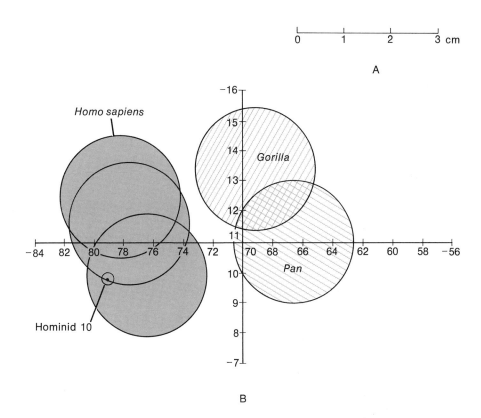

A

B

of mosaic evolution. The leg bones come from a hominid with relatively elongated lower limbs, as in man, and indicate a height of approximately 4 to $4\frac{1}{2}$ feet. As emphasized before, *A. africanus* and *A. habilis* were of similar body size, and both were smaller than *A. boisei.*

Higher up in Bed I, close to the top, the Leakeys' expedition discovered the terminal phalanx of a hominid big toe (Hominid 10). Day (1967) has analyzed this bone, again using canonical analysis. Selecting nine measurements, angles, and indexes that reflected functionally significant characters, he was able to distinguish clearly between men and African apes. Olduvai Hominid 10 fell within the human population range and therefore comes from a bipedal hominid. Size criteria suggest that this was more likely *A. habilis* than *A. robustus*, although this is by no means certain.

Still further up the section, from the base of Bed II, has come the badly crushed skull and dentition of a young adult hominid (Hominid 16) (Leakey, 1966). The cranium has been reconstructed by Dr. Mary Leakey and Professor Tobias; the estimated (and approximate) brain volume was something over 600 cm³. The skull has rather more prominent brow ridges and is relatively a little longer than that of *A. africanus*, in these features approaching *H. erectus*. However, unlike *H. erectus* and like *A. africanus* the skull bones are thin. The dentition shows minor differences from *A. habilis*, but is basically similar. Both Leakey and Tobias (personal communication) believe that Hominid 16 is not an *A. habilis;* Leakey (1966) regards it as a proto-*H. erectus* and believes that *A. habilis* only is ancestral to modern man. I think that it is quite probable that Hominid 16 represents a male of the *A. habilis* lineage.

More members of the same lineage have been found a little above the faunal change in Bed II (Hominids 13 and 14). The first of these (Hominid 13) is a juvenile individual with most of the skull and upper and lower dentition and mandible. This specimen may be in the region of $\frac{3}{4}$ million or more years younger than the oldest Olduvai specimens of *A. habilis.* The skull is gracile with only slight muscular markings, and it is probably a female. Holloway (1970) has estimated the brain volume at around 600 cm³. The skull is thin walled and is similar in a number of ways to *A. africanus*, although it is more advanced in some features. For example, the mandibular fossa (the joint for articulation of the lower jaw) is intermediate in morphology between *A. africanus* and *H. erectus* (Martyn and Tobias, 1967).

The dentition is very similar morphologically to the *A. habilis* type (Hominid 7), although some 20 per cent smaller. This may simply

Figure 72 Mandible of *Australopithecus habilis* (Homonid 13) from Olduvai Gorge Bed II (left). On the right is a hominid mandible from earlier Pleistocene deposits in Java. (Courtesy of P. V. Tobias.)

be due to sexual dimorphism, or it might indicate that the dentition was becoming reduced in this lineage. Although similar in overall size to *H. erectus* teeth, certain morphological features differentiate the two. The mandible is less massive in Hominid 13 than in other *Australopithecus* (including Hominid 7), a feature probably correlated with smaller tooth roots, and this has altered the internal mandibular contour. This hominid is quite clearly part of the same lineage as Hominid 7, yet shows some features of resemblance to later hominids. This is to be expected from its intermediate temporal position and should not cause undue taxonomic concern.

Thus we can trace at Olduvai a lineage of small hominids from the early Pleistocene through to the middle Pleistocene, one that shows some evolutionary changes during this time period. Postcranially it seems to have been very similar, in size as well as functionally, to *A. africanus*. The dentition is a little different from that of the latter species and the brain somewhat expanded.

The mean brain volume of *A. africanus* was around 440 cm^3, and the three *A. habilis* specimens yield estimates of some 600 cm^3 or more (Holloway, 1970). This is an increase of over a third, probably with little change in body size. If we allow a mean age of $2\frac{3}{4}$ million years for *A. africanus* and $1\frac{3}{4}$ million years for *A. habilis* (and if we further assume that one was ancestral to the other in the sense that South African *A. africanus* samples are broadly representative of their supposed contemporaries in East Africa), then there was an increase in brain volume of between 40 and 50 per cent per million years. Tobias (1967) has estimated the mean brain volume of *H. erectus* at around 940 cm^3. If we assume a mean age for that species of 1 million years, then between *A. habilis* and *H. erectus* the brain expanded at the rate of about 75 per cent per million years. There was also an increase in body size during this time, but it seems possible that relative brain expansion accelerated during the middle Pleistocene. The extent to which the brain of *A. africanus* had expanded compared with that of its ancestors is almost impossible to assess. Brain volumes for the pygmy chimpanzee (of approximately similar body size to *A. africanus*) are in the region of 300 cm^3, and such volumes may well have been typical of early Pliocene hominids such as *Ramapithecus*, although this is pure conjecture. In that case, there was something like a 50 per cent increase over an approximately 7 million to 10 million year period, probably not distributed at an even rate. Possibly the acceleration started with the neuroanatomical and behavioral changes associated with language and tool making.

A. habilis was apparently closer to *A. africanus* than to *H. erectus* in brain volume, other cranial features, dental and mandibular characteristics, and postcranial anatomy. It is normal in mammalian taxonomy for species within a genus to be similar in postcranial adaptations. For all these reasons, I prefer not to place *A. habilis* in *Homo*. It could be placed in a genus intermediate between *Australopithecus* and *Homo*, but this is almost certainly unnecessary. We should of course guard against the feeling that the (arbitrary)

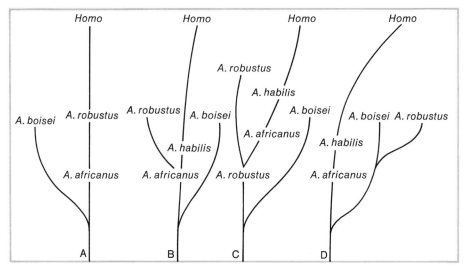

Figure 73 Alternative phylogenies for *Australopithecus* species. The author favors D. See note on page 156. (Pilbeam.)

"specific" boundary between *A. africanus* and *A. habilis* is in any way different from the "generic" distinction between *A. habilis* and *H. erectus* (Simons, Pilbeam, and Ettel, 1969).

Stone tools have been found throughout the Olduvai sequence, both scattered and on living floors. The living floors contain remains of butchered animals. At one lakeside site, low in Bed I, an apparently artificial circle of stones suggests that some kind of windbreak had been erected (Leakey, 1967). Thus early hominids were making a wide range of stone tools for the collection and preparation of a variety of animal and vegatable foods. They were certainly hunters (and this implies a division of labor between males and females), and they probably ranged over a wide area. Unlike the apes, they appear to have stayed at one camp for periods of days or even weeks and may well have returned year after year to the same sites. Thus in locomotor and behavioral terms they were much more similar to man than to apes. Certainly these hominids had long since crossed the threshold from apedom.

There is no way at present of telling whether or not both Olduvai hominids made stone tools. It is probable that *A. boisei* was less carnivorous than *A. habilis* and for that reason may have had less use for meat-preparing tools, but this does not mean that *A. boisei* would not have needed tools.

Before moving on to other areas, we should take note of the relationship between *A. habilis* and *"Telanthropus"* from Swartkrans. In their original description of *A. habilis*, Leakey, Tobias, and Napier (1964) mentioned dental and mandibular resemblances between *A. habilis* and the Swartkrans hominid, and Tobias and Professor von Koenigswald (1964) have also discussed similarities specifically

THE ASCENT OF MAN

between Hominid 13 and *"Telanthropus"* (if the latter is indeed distinct from *A. robustus*). Swartkrans is probably stratigraphically older than upper Bed I and lower Bed II at Olduvai. Should these East and South African hominids prove to be conspecific, this would pose nomenclatural problems, for the name *A. capensis* would then have "priority" over *A. habilis*. However, the Swartkrans material is probably too fragmentary to warrant such a change, although it does seem possible that the same late Pliocene or early Pleistocene gracile hominid species is being sampled in both South and East Africa.

OMO

In 1933, Professor Camille Arambourg collected specimens from fossiliferous locations in the Omo Basin of Ethiopia, North of Lake Rudolf. These deposits were thought, until recently, to be approximately equivalent in age to the lower parts of the Olduvai sequence. Since 1967 an international expedition involving contingents from the United States, France, and Kenya has been exploring this area. The United States group, led by Professor F. Clark Howell, and the French contingent have recovered approximately 100 hominid teeth, four jaws, and a partial skull (Howell 1968, 1969).

The amount of time spanned by the Omo Beds is now known to be much greater than was previously thought. A series of radiometric dates has been obtained for tuffs throughout the section, and it is now known that these sediments range in age from around $1\frac{3}{4}$ million years to more than 4 million years. Thus these Omo Beds finish where Olduvai begins, and the hominids from Omo enable us almost to double our knowledge of human prehistory in East Africa.

Until 1970 no definite stone tools were known from Omo, but in that season both parties of the expedition located Oldowan tool occurrences in deposits between 1.9 million and 2.2 million years old, and tools have also been found by Richard Leakey in Northwest Kenya east of Lake Rudolf in deposits of greater age (2.6 million years) (Howell, personal communication; Leakey, 1970b). Thus the origins of tool making have been pushed back almost a million years before their first appearance at Olduvai.

Howell (1969) believes that he has at least two hominid lineages represented at Omo. One is represented by a piece of mandible— lacking ascending rami—containing most of the dentition (number L7-125). The jaw bone is extremely massive and thick and strongly buttressed anteriorly. The cheek teeth are very large, the canines and incisors relatively tiny. Premolars especially, and molars too, are extremely broad—presumably to increase chewing area without facial lengthening. In all features this mandible is compatible with the upper teeth known from *A. boisei* (Olduvai Hominid 5), and

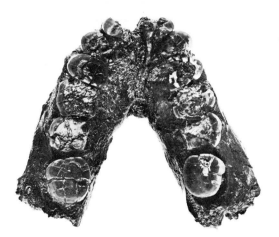

Figure 74 Occlusal view of *Australopithecus boisei* mandible from Omo (Courtesy of F. C. Howell.)

it has been tentatively referred by Howell to that species. The first two molars are particularly heavily worn, their enamel remaining only as a marginal ring surrounding a dentine basin, and the cheek teeth are greatly compressed one against the other. Both features indicate strong masticatory stresses. The third molar crown has been worn flat without exposing dentine, showing that the projecting cusps on hominid teeth are composed almost entirely of enamel. Intercuspal relations appear to be unimportant in guiding tooth and jaw movements in hominids; rather, the flattened opposing tooth surfaces can be sheared across each other in almost any direction. The type of tooth wear (similar to that in geladas, *Gigantopithecus,* and *Ramapithecus,* as well as in other *Australopithecus*) and the proportions of front and cheek teeth once again indicate a basically herbivorous diet of numerous small tough morsels, requiring powerful chewing.

The same lineage is probably represented by a mandible with tooth roots only discovered in 1967 by the French party at a horizon dated at more than $2\frac{1}{2}$ million years old (Arambourg and Coppens, 1968). Although smaller than L7-125, the mandible is of basically similar type. It was described originally as *Paraustralopithecus aethiopicus,* although this name has no biological validity. Possibly the differences between the two mandibles are due to sexual dimorphism or more likely to an increase in size through time. Another half mandible plus some isolated teeth from a horizon stratigraphically intermediate between these two mandibles fills in the morphological gap between them. Several other isolated teeth from various levels in the section, the oldest having an age of more than 3 million years, appear to belong to the same lineage. Finally, Richard Leakey's expedition to East Rudolf has recovered an almost complete *A. boisei* skull, lacking the upper teeth, from deposits at Koobi Fora dated at over 2 million years (Leakey, 1970a,b).

Thus the lineage which includes the type of *A. boisei* (Olduvai Hominid 5) and the Omo mandible (L7-125) has a considerable time range in East Africa, from more than 3 million years to around $1\frac{1}{2}$ million years ago or less. There is some evidence to suggest that during this period the lineage became dentally more massive, and this was probably correlated with changes in cranial robusticity, too. Whether or not the entire lineage should be described as *A. boisei* is an open question. It is likely that robust australopithecine sites from South Africa are over 2 million years old; earlier parts of the lineage should probably be called *A. robustus.*

Howell (1969) has evidence of another hominid at Omo, one with much smaller and more gracile teeth. This species is known mostly from a number of isolated upper and lower cheek teeth, which are distributed throughout the section in deposits ranging in age from more than 3 million to around 2 million years. In certain features these teeth show strong resemblances to those of Olduvai Bed I *A. habilis.* For example, an isolated second lower premolar (W-23) is elongated and narrow in a manner very similar to that tooth in Hominid 7 from Olduvai. It seems highly probable that the Omo teeth are sampled from an earlier part of the same East African lineage known at Olduvai as *A. habilis.*

The gracile Omo teeth also exhibit marked similarities to those of South African *A. africanus,* particularly the Sterkfontein sample. As noted before, it is likely that Sterkfontein is around 2 million to 3 million years old. Howell has classified the Omo teeth tentatively with *A. africanus,* and it does seem quite probable that Omo and Sterkfontein are samples drawn from approximately contemporaneous but widely separated subspecies within *A. africanus.* Parts of a juvenile skull from approximately $2\frac{1}{2}$ million-year-old deposits have been recently found at Omo and may belong to *A. africanus;* the brain size is smaller than *A. habilis* volumes and similar to specimens from Sterkfontein. Richard Leakey (1970b) has reported a similar skull from Koobi Fora in Northwest Kenya, dated at 2.6 million years. These provide confirmatory evidence that *A. africanus*

Figure 75   Occlusal view of robust *Australopithecus* hemimandible from Omo. (Courtesy of F. C. Howell.)

was ancestral to *A. habilis*. That is, these "species" are merely different parts of a single evolving lineage changing through time principally in brain size and tooth structure.

A third hominid species may be represented at Omo by a half mandible (L74-21) from the same levels as the *A. boisei* lower jaw (Howell, 1969). Only the canine and second premolar crowns are preserved, although the proportions of the front premolar can be inferred from contact facets on the adjacent teeth. The mandible is robust, and the cheek teeth are relatively large. The second premolar is very molarized and is almost square. It is rather different from its homologue in *A. boisei* and *A. africanus*, and it also differs to a lesser extent from *A. robustus*. Even more distinctive is the great length of the first premolar; it is usual in *Australopithecus* species with large second premolars for the first to be the smaller of the two. In the half mandible the front premolar was clearly heavily molarized. In other mandibular features L74-21 resembles robust *Australopithecus* species, particularly in the anterior buttressing as well as the vertically oriented and forwardly placed ascending ramus. The taxonomic status of this jaw remains uncertain at present; probably it lies in the robust lineage.

The habitat at Omo was probably relatively open woodland and parkland, close to the shores of a meandering river (Howell, personal communication). Evidently, hominids were occupying living sites in this area as long ago as $2\frac{3}{4}$ million years. Which of the Omo hominids was (or were) responsible for the stone tools is as much open to debate as the identity of the Olduvai tool maker. On balance, the gracile lineage (*A. africanus–A. habilis*) is very likely to have manufactured and utilized implements; however, the other species may also have been making tools. At such a primitive level of technology it will probably be difficult for archeologists to differentiate between tool types made by different early hominids. Thus the Oldowan culture may in fact be a technological complex of almost indistinguishable industries.

PENINJ

Early in 1964, a worker on an expedition to Lake Natron in Tanzania led by Dr. Glynn Isaac and Richard Leakey discovered an almost complete hominid mandible, with all the teeth preserved, in beds close to the Peninj River. The dating of this mandible has proved possible, for the associated fauna and early Acheulean tools are equivalent to Olduvai upper Bed II, and a radiometric analysis of a basalt just above the jaw gave a date of about $1\frac{1}{2}$ million years; this may well prove to be the age of Bed II at Olduvai. The dating of the Peninj jaw has therefore helped to clarify the Olduvai situation (Isaac, 1967, 1969).

The mandible has yet to be described thoroughly. Some workers have regarded it as *A. boisei* (Leakey and Leakey, 1964), others as

Figure 76 Occlusal view of probable *Australopithecus robustus* mandible from Peninj. (Pilbeam.)

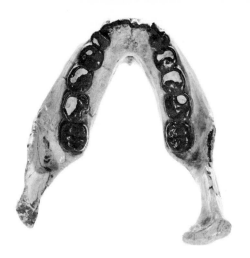

A. robustus (Tobias, 1967). It does not completely resemble either, although it is closer to *A. robustus* in a number of features in which it differs from *A. boisei*—for example, in the height of the ascending ramus and in premolar morphology. It could be sampled from the same lineage as *A. robustus* and *A. boisei*.

The half mandible from Omo (L74-21) also shares some features with the Peninj jaw and may be sampled from the same lineage. At present, little more need be said, except that it is a robust *Australopithecus*.

GARUSI

In 1939, Kohl-Larsen discovered some fragmentary hominid remains from the Laetolil Beds at Garusi, northwest of Lake Eyasi in Tanzania. These deposits are probably approximately equivalent in age to the lower part of the Omo section (around 3 million years). Kohl-Larsen found a fragment of maxilla containing both upper premolars; some distance away he recovered the worn upper third molar of another individual. These he assigned to *Meganthropus africanus*, a biologically nonvalid species. Robinson (1953) has shown quite clearly the considerable similarity between the Garusi premolars and South African *A. africanus*, and this also applies to the more recent sample from Omo. Therefore, we can assume that this fragment probably represents *A. africanus*.

KANAPOI

Kanapoi is a site in Northwest Kenya near the southern end of
Lake Rudolf. Professor Bryan Patterson discovered a hominid distal
humerus there in 1965 from beds that are capped by lava dated
to be $2\frac{1}{2}$ million to 3 million years old (Patterson and Howells, 1967).
The fossiliferous locality could be up to $4\frac{1}{2}$ million years old. Patterson and Professor W. W. Howells (1967) have compared the
distal humeri of man and the chimpanzee (the hominoid most
similar to man) using a multivariate statistical technique known
as discriminant function analysis and conclude that the Kanapoi
humerus is closely similar to that of modern man and very different
from that of the chimpanzee. It differs also from the Kromdraai
A. robustus humeral fragment in a few features.

Presumably, these resemblances and differences indicate that the
Kanapoi forelimb was not used to support the body's weight as
in knuckle walkers but rather in a much more human way. Patterson and Howells believe that the humerus may well be from an
A. africanus; they point out that they have examined human humeri
that match almost exactly the Sterkfontein fragment proximally and
the Kanapoi specimen distally. If this is the case, it pushes the
known remains of the *A. africanus–A. habilis* lineage still further
back. However, the humeral fragment is large and it is more
reasonably assigned to *A. boisei.*

LOTHAGAM

Patterson (Patterson, Behrensmeyer, and Sill, 1970) has discovered
a still older Kenyan hominid at Lothagam. The site has a probable
age of around 5 million years. The specimen has not yet been
described; it is a piece of mandible with the first molar crown
preserved and the roots of the other molars. In crown pattern, size,
and proportions, and in other mandibular features, too, the specimen is very similar to *A. africanus* from Sterkfontein and
Makapansgat. It possibly represents *A. africanus* and would extend
the time range of that species still further back in time.

BARINGO

Recent fieldwork in the Baringo Basin in the Northern Rift Valley
of Kenya under the general direction of Professor B. C. King and
Dr. W. W. Bishop (personal communication) has resulted in the
recognition of sediments ranging in age from late Pleistocene to

Figure 77 *Australopithecus africanus* from Makapans (left) and the mandibular fragment, possibly of the same species, from Lothagam (right). (Courtesy of B. Patterson.)

late Miocene. This is a potentially very exciting area, for it includes deposits that span the time between the youngest *Ramapithecus* and the oldest *Australopithecus.*

Two hominid fossils have been recovered from these beds. The first comes from the upper Fish Beds of the Chemeron Beds (Martyn and Tobias, 1967). An estimated age for these deposits is 3 million to $3\frac{1}{2}$ million years. The specimen is most of a right temporal bone, and this has recently been described in detail by Tobias, who has not assigned it to a species but merely discusses its resemblances to and differences from the various *Australopithecus* species.

The second specimen, a second upper molar, comes from the Ngorora Beds, around 9 million years old (Bishop and Chapman, 1970). This is one of the most exciting and yet tantalizing finds of the last few years, exciting because it fills in the gap between *Ramapithecus* and *Australopithecus,* tantalizing because it is so fragmentary. The cusp form and general distribution of crests, grooves, and so forth are more reminiscent of *Australopithecus* than of the earlier hominid, yet it is low-crowned like that of *Ramapithecus.* It is thus in a number of ways intermediate both in time and morphology between these two major groups of hominids.

The hominids appear to have evolved as forest or forest-fringe animals feeding in open spaces in wooded areas. During the Pliocene a general cooling of Earth's atmosphere combined with increasingly seasonal climates seems to have reduced the amount of evergreen forest and replaced it in many areas by deciduous woodland and tree savanna. *Ramapithecus* of the late Miocene and early Pliocene was a forest dweller; by the middle Pliocene, descendants (probably) of *Ramapithecus* had become adapted to life in more open country. *A. africanus* is known from East and South African middle and late Pliocene sediments and quite possibly was distributed elsewhere in Africa—perhaps even in India, too—at this time. *A. africanus* was probably descended from *Ramapithecus;* the tantalizing Ngorora tooth from Baringo certainly suggests a tie-up between the two. Along with changes in habitat, the *Ramapithecus-Australopithecus* transition was accompanied by the following changes (some more speculative than others).

The brain increased in size, probably as a result of internal reorganization associated with basic behavioral changes. These changes are most likely to have resulted in a brain capable of language production and tool making, the type of brain that could impose arbitrary form upon its environment, thus generating at least the beginnings of what was to become human culture. The selective pressures that caused these changes are as yet unknown.

Hominids became bipedal, resulting in changes in postcranial anatomy and function. These changes also involved brain reorganization, particularly in the cerebellum. Forelimbs once freed from locomotor activities could indulge in ever-increasing amounts of finely controlled manipulative behavior that in turn involved brain changes. The selection pressures producing these locomotor changes were probably complex. Primates are bipedal in a variety of situations: for example, while carrying objects, particularly food, or when scanning their surroundings for predators. The African apes are also bipedal during display behavior, when excited or aggressive. Their displays at these times may include the waving and throwing of objects—branches or stones. If such behaviors were part of the repertoire of *Ramapithecus,* then during the Pliocene selection for hunting may well have favored the development of bipedalism.

At least some species of *Australopithecus* were hunter–gatherers; judging from the diet of living peoples such as Bushmen and Australian Aborigines, upward of one third of the food probably came from hunting, the remainder from plants. The adoption of hunting would have resulted in a number of profound behavioral changes. First, each troop would range over a much wider area than is typical of vegetarian foraging primates. The woodland chimpanzees of Western Tanzania have the largest home ranges of the nonhuman primates, a troop of fifty or so animals ranging through an area of up to 80 square miles. Probably *Australopithecus* home-range size was even greater, and population density consequently lower, as the shift to hunting occurred. Bipedalism would facilitate traveling over longer distances and permit weapon carrying and also, perhaps most important, weapon use. Transporting game back to camp, together with raw materials for tool making, would become much easier for a biped.

The adoption of hunting would have produced for the first time in primate history a division of labor between the sexes. The females retained their roles as gatherers of vegetable food (also, of course, they were responsible for care of young), while the males were hunting. For hominids, hunting, as for the small carnivores such as wolves and hunting dogs, requires cooperation among a number of individuals. Not only would the hominid troop be split up when most of the adult males went on hunting expeditions, but for the first time cooperative behavior among males (and also of course among females) would become essential. As a further corollary, the shift to cooperative hunting would of necessity require food sharing between males and females and probably also among subgroups within the troop. Both cooperative behavior and food sharing are atypical behavior in nonhuman primates, occurring only occasionally in the chimpanzee.

With the males away on periodic hunting expeditions, hominid troops would have ceased moving their sleeping area each night as the vegetarian foraging primates do but instead would have set up temporary campsites for periods of several days. Thus there was a base camp for the males to return to with game and a place for butchering, sharing, and eating the meat over a period of days.

The importance of hunting has perhaps been overemphasized in discussions of hominid evolution; undoubtedly, many of the stone implements used by hominids were meat-preparing tools, yet the technology involved in plant collection and particularly in its preparation was perhaps equally important. Much of the plant food available in open country requires quite complex prepara-tion—crushing, soaking, and so forth—and this aspect of the hominid's developing technological skills should not be overlooked.

The growth of cooperative behavior among members of the troop would have produced a change in dominance interactions, particu-larly among males. This required not only suppression or control of aggressive behavior but also the development of new neural structures for positively mediating cooperative activities. The re-duction of male aggression would have facilitated their integration into the troop in a positive way, perhaps through the development of permanent pair bonds between male and female.

Perhaps the most important change of all would have been the evolution of language. All primates communicate vocally as well as by gestures and postures, but what they communicate to others involves emotion—fear, alarm, and so forth. Human language com-municates much more, particularly abstract concepts involving removal in time and space that must have been important for the early hominids, and requires the ability to name objects, and to organize discrete sounds in such an orderly way as to convey meaning. Hominids evolved new interconnections within the brain for language production, and these changes may already have occurred by the late Pliocene. Language, once evolved, would have conveyed considerable selective advantages on those hominids possessing it. Cooperative hunting required planning, both among the males and also between the males and the females and young, who remained behind.

Planning and forethought, cooperation, and memory are all hominid behavioral characteristics essential to our success, and it seems highly likely that it was the transition to the hunting-gathering way of life in relatively open country that produced these changes. Certainly the hunting aspects of this new way of life would have required all these traits, but the gathering part should not be forgotten. Gathering enough plant food in savanna or savanna woodland subject to a seasonal climate would also require skill, knowledge of the terrain, planning, and memory.

It is possible that during this period male–female permanent bonds were evolved. This type of social bond is very uncommon among primates—the gibbon is the only Old World form in which it occurs—for they normally live in multimale groups in which mating is more or less promiscuous ("more or less" indicating that there is growing evidence that mothers and sons almost never mate). The most stable units within primate troops are mother centered— adult females with their offspring. Ties between mothers and children are maintained long after the children attain adulthood. Probably, human nuclear families were formed by the addition of a male to these female-oriented groupings, presumably because it was essential for the female-offspring unit to have a reliable and predictable provider.

The evolution of permanent bondings probably resulted in the development of incest taboos tetween fathers and daughters and between brothers and sisters as well as between mothers and sons. The application of these "rules" would have become much easier after the evolution of language had made possible the codifying of such concepts as "mother," "father," and so forth. It is also quite likely that some form of exogamy was practiced. At adolescence many primate males apparently leave the troop into which they were born to join another where they often remain throughout their adult life. This sort of pattern may well have been characteristic of the early stages of hominid evolution, although with the development of hunting it is just as likely that females rather than males would have been exchanged, for males would need to know well other males with whom they hunted and to be familiar with their hunting territory. In this way, a spreading network of political ties would be built up, acting to reduce potential sources of tension between troops.

Another change that can be inferred is a shift in female sexual behavior from cyclical to continuous receptivity. All female primates other than human females are available as sexual partners only at estrus, a short period before and after ovulation. Human sexual receptivity throughout the cycle probably served to strengthen the pair bond.

The evidence from the maturation of *Australopithecus* jaws and teeth suggests that infant and juvenile periods of life were more extended than in apes, implying longer maternal (and paternal) ties with offspring and in turn indicating that the amount of time available for learning socially important skills was increased. If these early forms really did possess even rudimentary language, then the opportunities for such learning would have been doubly enhanced and the lengthened maturation explained.

So it is likely that by the end of the Pliocene there emerged a totally new form of hominid behavior based on hunting and involving linguistic communication, tool making, division of labor, pair bonding, and prolonged contact with infants, and requiring new behavior patterns based on cooperation, delayed gratification, new forms of motivation, reduction of interindividual aggression, and so forth. One further point about language: the development of this form of communication would have made possible for the first time in primates the reward and reinforcement of nonaggressive behavior patterns. Dominance ceased to be its own reward.

The other important change that had definitely occurred by the terminal Pliocene and that seems to have been well under way even 10 million years before was the evolution of the characteristic hominid dentition. This apparently evolved in response to a new type of masticatory behavior, powerful slicing anteriorly and crushing and grinding posteriorly. The result was the reduction and incorporation of the canines into the incisor cutting battery and the development of thick-enameled flat-crowned cheek teeth that could be used in chewing in almost any combination of movements. For some time it has been assumed that the reduction of male hominid canine teeth was correlated with their replacement by hand-held weapons. The reason for the reduction of male canines so early in hominid evolution and for the radical change in structure and function in becoming part of a completely new masticatory pattern was basically (at least in the beginning) a herbivorous adaptation to feeding in an open-country niche. The exact nature of the plant food eaten and the way in which it was eaten are unknown at present.

The pattern of hominid evolution in the late Pliocene and early Pleistocene is now fairly well established. At least two lineages were present in East Africa, *A. africanus* and *A. robustus/boisei;* the former also lived in South Africa at that time ($2\frac{1}{2}$ to 3 million years ago). There is some evidence that indicates that the robust lineage became somewhat more robust during this period. *A. africanus* evolved by the early Pleistocene into *A. habilis* in East Africa (possibly in South Africa, too). It is possible that southern populations of *A. africanus* produced the later *A. robustus,* although this species seems more likely to be related to *A. boisei.* The original time of speciation of *A. africanus* and *A. robustus* is unknown but may well have been middle Pliocene. The common ancestor of all *Australopithecus* was probably most like *A. africanus* in skull structure and function and tooth proportions.

A functional interpretation of these early hominids suggests that cranial and mandibular morphology is closely correlated with masticatory function and tooth size. Thus the various species are essentially "variations on a theme," and so the conversion of one into another can be seen as an essentially simple genetic process, chanelled and controlled by selective pressures associated primarily with feeding behavior, particularly the type of food eaten.

The Pliocene saw the emergence of bipedal, cooperative hunting hominids and ushered in a unique phase in primate evolution. The basic hominid behavior patterns were well established by the end of the early Pleistocene, at which time a new type of human ancestor emerged, *Homo erectus.* Let us turn now to a consideration of these first men.

Note added in final proof stage of this book: Recent work by Cooke (personal communication) suggests that South African *Australopithecus* sites are between 2 and 3 million years old. The probable redating of *A. robustus*, together with new material uncovered at East Rudolf by Richard Leakey (personal communication), suggests that robust australopithecines fall into one lineage, the earlier part of which should be classified as *A. robustus*, the latter part *A. boisei*. Accordingly, Figures 6 (page 13), 64 (page 129), and 73 (page 144) should be modified.

Middle Pleistocene Hominids

7

The Pliocene and earlier Pleistocene hominids appear to have been confined mainly to Africa, although this conclusion is based on negative evidence, because Arabia, Europe, India, and other parts of Asia have yet to be adequately surveyed. There is some evidence to suggest that hominids had spread at least to Java by the end of the Pliocene (Curtis, personal communication). However, during the middle Pleistocene, commencing by at least 1 million years ago, hominids or their cultural remains were definitely distributed throughout much of the tropical and subtropical parts of the Old World. The hominids of this time period, between about $1\frac{1}{2}$ million and $\frac{1}{2}$ million years ago, show a number of significant advances over *Australopithecus*. They were taller, had larger brains and relatively smaller faces and teeth, and probably were capable of more complex behavior. Hominids of this period are generally classified as *Homo erectus* (Howells, 1966).

The first *H. erectus* were found in what is now Indonesia (Java) in the nineteenth century, and the largest sample of this species was recovered from Northern China mainly in the 1920s and 1930s. Middle Pleistocene hominids from elsewhere in the world were practically unknown during this time—almost certainly because of sampling, because suitable sites for the preservation of middle Pleistocene hominids in Europe and Africa had not been located or thoroughly excavated. Thus the idea developed that *H. erectus* was an Asiatic species. When forms attributable to the species were discovered in other places, the tacit assumption has sometimes been made that they had "migrated" into the area from the East. This is almost certainly not true. As far as we can tell, hominids were quite widely distributed throughout the Old World at least by the earlier parts of the middle Pleistocene ($1\frac{1}{2}$ million to 1 million years ago) and probably even earlier; the distribution of known fossils is a function of the position of sites of suitable age and the exhaustiveness of excavation, not of hominid distribution. Although this is a controversial point, it seems highly likely that, unlike the Pliocene-Pleistocene hominids, human forerunners in the middle Pleistocene belonged to one species. This species, *H. erectus*, shows considerable regional variation in various morphological features between populations. These interpopulation differences are similar in kind and degree to the subspecific ("racial") differences between groups within any wide-ranging mammal species (including *H. sapiens*) (Coon, 1962).

In a general sense, *H. erectus* appears to have been ancestral to the archaic groups of man, some of which in turn were probably ancestral to modern human populations. In a number of regions, there is considerable continuity between populations ranging in age from early middle Pleistocene to late Pleistocene, populations that probably represent rather coherent (through time) infraspecific lineages. We can envisage human evolution during the second half of the Pleistocene as an (Old) world-wide ongoing process; at any given time level, only one species seems to have been present— although the populations comprising this species sometimes show considerable regional variation—and this species changed through time. For convenience, the earlier parts of the lineage are called *H. erectus* and the later parts, because so much change has occurred, *H. sapiens.* Yet it should be emphasized that we are dealing with a continuum, and therefore some temporally intermediate populations will be difficult to assign to either "species."

Not only are geographically separated contemporaneous populations related, but within broad regions populations are interconnected through time. Thus middle Pleistocene *H. erectus* populations in Asia and Africa are similar because they belong to the same species, whereas African *H. erectus* and late Pleistocene *H. sapiens* may also show similarities because they are sampled from a relatively geographically stable lineage within the evolving species. All these factors have to be taken into account in evaluating the position of fossils. We have relatively few samples from widely scattered parts of the Old World to fill in a vast amount of time. Inevitably there is a great deal of extrapolation involved, and the hypothesis that I prefer here is not by any means the only possible way in which the fossil facts can be organized.

We shall discuss these middle Pleistocene hominids area by area, and do so more or less chronologically in terms of their discovery.

INDONESIA

Fossiliferous hominid-bearing deposits of later Pleistocene age were first discovered near Trinil in Eastern Java in the 1890s (Dubois, 1894, 1924). Since then other sites have been located in this area and seem collectively to span the last 2 million years. Three characteristic faunal assemblages have been defined in Java, the Djetis, Trinil, and Ngandong faunas, ranging from oldest to youngest (Hooijer, 1962). The ages of these faunas have been debated vigorously, and inconclusively. Recently though, some radiometric ages have become available. A basalt from a formation presumed to correlate with upper parts of deposits of Trinil age has yielded an age estimate of $\frac{1}{2}$ million years, and tektites (tiny meteoritic glasses) from the same formations have been dated at around 700,000 years (von Koenigswald, 1962). Other radiometric dates indicate even older ages (Curtis, personal communication). Thus the Trinil beds apparently range in age from around 1 million to less than $\frac{1}{2}$ million years. The underlying Djetis levels probably run back to at least 2 million years ago, perhaps more. Thus the Djetis and Trinil faunas are correlative with Beds I to IV at Olduvai. The Ngandong faunas are younger, late Pleistocene in age, and their absolute date is unknown. They are probably less than $\frac{1}{4}$ million years old.

The first *H. erectus* was discovered by Dubois in 1891 near Trinil and consisted of a brain case, lacking the cranial base. In the following year, Dubois found a femur in the same beds, which he assumed belonged with the skull. These bones were classified as *Pithecanthropus erectus* and were later transferred to *Homo*. The fragmentary cranium has large brow ridges, a generally flattened appearance with markedly projecting occipital region, and a cranial capacity of 850 cm^3 (Le Gros Clark, 1964). The bones of the vault are very thick, much more so than in either *Australopithecus* or *H. sapiens* (Tobias, 1967). In contrast to the primitive appearance of the skull, the femur closely resembles that of modern man. The relative size of the head, the length and orientation of the neck, the size and orientation of the trochanters, and the general robusticity of the shaft are very similar to *H. sapiens* and differ in these features from femurs of *Australopithecus* (Le Gros Clark, 1964). Clearly, as far as can be determined, the femur is that of an upright biped as fully adapted as *H. sapiens*. There was at first some reluctance shown by anthropologists to accept the skull and femur as belonging to the same species (a similar problem arose with *Australopithecus*), for earlier workers were not familar with the

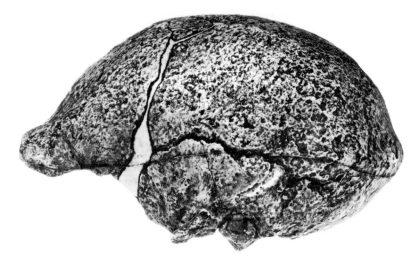

Figure 78 Side view of *Homo erectus* skullcap (cast) from the Trinil layers in Java (cast). (Courtesy of Wenner-Gren Foundation.)

concept of mosaic evolution. However, Dubois seems to have been a careful collector, and it is quite possible that the skull and femur came from the same deposits. Also, the chemical composition of both bones is similar to other elements of the Trinil fauna. Finally, remains of five other femurs were recovered later by Dubois from these sediments. (Recently, Day (*Nature,* 1971) has thrown doubt on these associations. He argues that material from Choukoutien constitutes the only known Asian *H. erectus* limb bones.)

Commencing in 1936, Ralph von Koenigswald made a series of discoveries of hominids in beds of both Trinil and Djetis age. A number of names have been applied to these specimens, but they all appear to be part of one lineage, extending through time.

Two additional skulls from Trinil horizons were discovered by von Koenigswald in 1937 and 1938 (von Koenigswald, 1940). Both are similar in size and form to Dubois' 1891 skull. They have endocranial volumes of 775 and 890 cm^3 respectively. A fourth skull from the Trinil layers was discovered in 1963 and reported by Sartano and by Jacob (1964) in the following year. Jacob estimates the volume at 975 cm^3, although this may well be an overestimate.

Three mandibles have come from the Trinil zones. Two of these are complete enough for evaluation. The teeth resemble in general those of *A. africanus,* although there are some more specialized features. However, the cheek teeth are rather smaller, and the incisors may have been a little larger. Tobias and von Koenigswald (1964) have recently pointed out the similarities between the dentition of *H. erectus* mandible B from Sangiran in the Trinil zone and the lower teeth of *A. habilis* Hominid 13 from Olduvai Bed II (see Figure 72). These specimens are from broadly the same time span, although the Indonesian jaw is probably younger. They are very similar, although the Olduvai specimen has narrower teeth. However, the skull of Hominid 13 differs in a number of features from *H. erectus,* and for this reason and others (for example, inferred body size), is retained here in the classification as *A. habilis.* Yet this is to some extent a matter of opinion, for the two samples appear to be drawn from a single evolving species lineage.

From the Djetis levels have come two skulls, so-called *H. erectus* IV found near Sangiran in 1939 and a child's skull from Modjokerto discovered 3 years before that. The adult skull consists of part of the face and upper jaws with teeth and most of the rear portion of the cranium. The brain volume has been estimated at 750 cm^3, the smallest known *H. erectus.* Its teeth have also been compared by Tobias and von Koenigswald (1964) with the uppers of Olduvai Hominid 13, and the resemblances are again marked, although

Figure 79 Femurs of *Homo erectus* (left) and *Homo sapiens* (right), showing the close similarity in morphology (casts). Recently it has been suggested that the femur from Java (left) is not *H. erectus.* (Pilbeam.)

H. erectus has broad teeth like those in the mandible. The infant skull—it is thought to represent an individual 2 to 3 years old—is almost complete, although it lacks the face. Although the skull is from a young individual, when compared with an infant *H. sapiens* the brow ridges are relatively prominent, the frontal region is starting to narrow, and the back of the skull is angulated. The adult brain volume would have been hardly more than 800 cm^3.

Mandibles from the Djetis layer number three and have all been called *Meganthropus palaeojavanicus,* apparently because of their large size (Weidenreich, 1945). However, the teeth are closely similar to those of other *H. erectus* in Indonesia, China, and elsewhere. Tobias and von Koenigswald (1964) compared the *"Meganthropus"* mandible and teeth (premolars and first molar) discovered in 1941 with the type of *A. habilis* (Olduvai Hominid 7). There are many similarities between the two (see Figure 68), although Hominid 7 has narrower teeth. The Djetis and Trinil mandibles and teeth are about as similar to one another as are Olduvai Hominids 7 and 13 to each other. The *"Meganthropus"* specimens may be large because they are males and also quite likely because they are geologically older, perhaps as much as 1½ million years (Curtis, personal communication), than Trinil *H. erecuts.* Once again, the general similarities between the jaws and teeth in East Africa and Indonesia emphasize the fact that we are probably dealing with samples drawn from one lineage. The large mandibles from Djetis levels may well represent *Australopithecus;* this can not be settled definitely until more associated jaws and crania are recovered.

One of the Djetis skulls (IV) is more like *H. erectus* than *A. habilis* and should probably be assigned to the former "species." The status of *"Meganthropus"* is equivocal. The Djetis deposits are very thick and cover a considerable span of time. The evidence suggests that hominids had spread out of Africa at least 2 million years ago, probably at the *Australopithecus* stage. The taxonomy of these Indonesian specimens is a little difficult. Clearly, they form a continuum that spans the *A. habilis–H. erectus* transition, and there seems little point at present in debating the position of the arbitrary boundary between these "species." The jaws and teeth of the Trinil and Djetis hominids apparently became smaller through time. The reasons for these changes are unknown but are possibly associated with improvements in food preparation techniques. There is also evidence that brain volume increased during this time period. The Djetis skulls are both 800 cm^3 or less in volume, whereas those from the Trinil horizons have an average of closer to 900 cm^3; however, the samples are so small that this apparent trend can only be considered a possibility at present (Tobias, 1967).

The differences between *A. habilis* and the earliest *H. erectus* can be summed up as follows. The brain became larger in *H. erectus* (from a mean of around 600 cm^3 to one of over 800 cm^3). The expansion was probably not general, because the skulls of *H. erectus,* although broader and longer than those of *A. habilis,* are barely higher. Expansion seems to have been in the parietal and temporal lobes, and possibly also in the cerebellum—and affected the overall shape of the skull, making it longer, broader, and flatter than in *A. habilis*. The causes of the brain changes are obscure, although they were almost certainly associated with behavioral changes. The bones of the skull vault are much thicker in *H. erectus* than in earlier hominids (and later ones, too). As will be seen when the Chinese forms are discussed, this bony thickness appears to have been a general species characteristic and not solely confined to the skull (Tobias, 1967).

Whereas *A. habilis* was a pygmy-sized creature, 4 feet to 4 feet 6 inches tall on average, *H. erectus* individuals were a foot taller; the middle Pleistocene species was also a fully adapted biped, unlike any *Australopithecus*. Selection pressures producing these changes were presumably associated with increasing group range size, mainly due to more efficient (perhaps big-game) hunting. The increase in body size probably contributed to at least some of the expansion in brain volume, although this is not the whole story, particularly not for later groups of *H. erectus* with still larger brains.

Nothing can be said about the social behavior of *H. erectus* in Indonesia, because the remains are scattered and isolated. No living floors are known, and there are no associated stone tools, although there are artifacts of equivalent age from other parts of Java (Oakley, 1961). The implements are rather crude and resemble the Oldowan tools in general level of sophistication.

Indonesia was an area of hominid occupation over a long period of time. The populations from the late Pleistocene Ngandong levels, to be discussed in the next chapter, show many similarities to Javan *H. erectus*, although they are in a number of ways more advanced. Possibly this peripheral area, in terms of Old World hominid evolution, was receiving some inflow of genes from other parts of Asia, but in general the later Pleistocene seems to have been a period of some stability for these hominids.

Let us now turn to middle Pleistocene hominids from other parts of the Old World.

CHINA

A few teeth from Chinese drugstores, collected by von Koenigswald (1957), may date from the early Pleistocene or early middle Pleistocene. He has described these forms as *Hemianthropus* (later *Hemanthropus*) *peii*, although they almost certainly do not represent a new genus and species. Other drugstore teeth have been described by him as *Sinanthropus officinalis*. Because the stratigraphic age of these specimens is so uncertain, little can be said about their taxonomy, although they do appear to lie in the general *Australopithecus–H. erectus* lineage.

The oldest undoubted *H. erectus* in China comes from Lantian

County, Shensi Province, Northern China. A mandible was discovered there in 1963 and parts of a cranium quite close by in 1964 (these do not belong to the same individual) (Woo, 1964, 1966). The associated fauna indicates a warm climate, and the deposit was probably laid down in rather open country. The ages of the sites are approximately equivalent to the Djetis or Trinil levels in Java (Chow, 1965), and the morphology of the hominids is also similar to that of the early Indonesian *H. erectus.*

The vault bones are thick, with massive brow ridges, flattened frontals, and an angulated occipital. What is preserved of the face suggests that it was more projecting than in *H. sapiens* (no faces are known from Indonesian *H. erectus*). The brain volume was under 800 cm^3, probably closer to 750 cm^3, thus confirming once again that the very earliest undoubted *H. erectus* had small brain volumes. The mandible lacked a chin, as did other early hominids; the teeth were relatively large compared with those of later men. The jaw has one peculiarity: congenital absence of the third molar, a condition that occurs with varying frequencies in modern man (particularly in Asian, American Indian, and Eskimo populations).

The largest known sample of *H. erectus* comes from Choukoutien near Peking (Weidenreich, 1936, 1937, 1941, 1943). The first specimen, a lower molar, was discovered and described in 1927 as *Sinanthropus pekinensis.* For the following 10 years the site was worked more or less continuously. Digging recommenced in 1949, and several more hominids have come to light since then (Coon, 1962). Regrettably, all but one specimen of those collected before the war were lost, apparently in China, after they had been packed to be sent for safekeeping to the United States at the beginning of the Japanese occupation of China. Fortunately, they had been exhaustively described in a number of monographs by Dr. Franz Weidenreich, and a series of good casts were made.

Choukoutien was apparently the site of fairly continuous hominid occupation for some time during the middle Pleistocene. The exact date of the deposits is unknown, but they are younger than the *H. erectus* discussed previously, perhaps between 800,000 and $\frac{1}{2}$ million years old. The climate fluctuated from cold to warm to cold, judging by the associated flora and fauna (Kurtèn, 1960). Occupation of the area was apparently only intermittent at first. A very large associated fauna has been recovered, and many of the bones are charred and splintered. Bigger game is well represented, particularly the remains of two species of deer. A number of hearths have been discovered, so *H. erectus* had invented the use of fire by this time, at least in more northerly latitudes. That he was not exclusively carnivorous is shown by the abundance of hackberry seeds recovered (Coon, 1962).

More than a dozen skulls, several mandibles, almost 150 teeth, and a number of postcranial bones constitute our knowledge of man at Choukoutien. The bones are peculiar in that all of them are thick, as in Indonesian *H. erectus.* The brain size ranged from around 850 to 1,300 cm^3 (the latter within the range of modern

Figure 80 *Homo erectus* from Choukoutien, China (cast). (Pilbeam.)

man), yielding a mean of over 1,000 cm^3. Thus the brain was larger than in the specimens from Java. This may be due to the younger geological age of the Chinese specimens, although other factors such as sampling bias and infraspecific population variation may be contributing factors. Although the brain expansion was probably general relative to the earlier forms, the parietal and frontal areas may have increased differentially a little. Thus the cranial vault is more rounded and higher, and has a less flattened frontal region. The face is still relatively large and projecting compared with that of *H. sapiens*.

Jaws and teeth show considerable size variation, possibly due to sexual dimorphism. In general, they are very similar to those of *H. erectus* from Java.

Weidenreich described parts of seven femurs, two humeri, a clavicle, and a carpal bone. Since 1949 another humerus and a tibia have been recovered (Coon, 1962). They are basically similar to those of modern man, although as noted the cortical bone is much thicker. The stature of Choukoutien man was only a little over 5 feet, shorter than the Indonesian forms. Short, stocky bodies are thought to be cold adapted, lowering the ratio of surface area to body volume and thus reducing heat loss, and this might be an important contributing factor in the small body size of the Choukoutien hominids.

Stone artifacts are scattered throughout the deposit and are rela-

Figure 81 Mandible of so-called Meganthropus from the Djetis layers in Java, compared with isolated teeth of *Homo erectus* from Choukoutien, China, showing their basic similarity (casts). (Pilbeam.)

tively crude, Oldowan-like tools (Oakley, 1961). No hand axes have been found; indeed, these tools seem to be confined to African, European, and West Asian sites during the middle Pleistocene.

These Chinese *H. erectus* were apparently behaviorally and technologically more advanced than earlier hominids. They were hunting big game in large amounts and probably ranging over a very wide area. Their use of fire and the charred bones mean that they were cooking their food. Dr. Edmund Leach (1967) believes that fire has uses other than for cooking and keeping warm:

> Now it isn't a biological necessity that you should cook your food, it is a custom, a symbolic act, a piece of magic which transforms the substance and removes the contamination of "otherness." Raw food is dirty and dangerous; cooked food is clean and safe. So already, even at the very beginning, man somehow saw himself as "other" than nature. The cooking of food is both an assertion of this otherness and a means of getting rid of the anxiety which otherness generates. (Leach, 1967:89)

Accordingly, we can conclude that important behavioral advances had been made by the hominids by the middle Pleistocene, particularly in the areas of linguistic communication and symbolizing.

A number of important middle Pleistocene hominids have been discovered in Africa during the past two decades. In 1954 and 1955, Arambourg (1963a) found three mandibles with teeth, a parietal bone, and two milk teeth at Ternifine near Oran in Algeria. The jaws resemble those of *H. erectus*, as do the teeth, and the skull fragment is very thick, Although Arambourg gave them a new name, *Atlanthropus mauritanicus*, these specimens almost certainly are *H. erectus*. Such differences from the Asian material as there are can be explained as racial variation. The Ternifine deposits are correlates of Olduvai upper Bed II and contain a hand-axe industry of early Acheulean type.

Somewhat later in time is the site at Sidi Abderrahman in Morocco from which have come two fragments of *H. erectus* mandible (Arambourg and Biberson, 1956). Still younger in age are the fragmentary remains of an adolescent hominid from Rabat, also in Morocco. This specimen is in many ways intermediate between North African *H. erectus* and archaic-man populations in the same area, implying population continuity in North Africa over a fair period of time.

Homo erectus remains are also known from Olduvai Gorge in Tanzania. In 1960, Dr. Leakey (1961) discovered a large hominid skull (Hominid 9) at a site high in Bed II, at the same level as the early Acheulean industries. The age of this specimen is probably around $\frac{3}{4}$ million years or more (Isaac, 1969). It has yet to be

Figure 82 Skullcap of *Homo erectus* from Olduvai Gorge, Tanzania (cast). (Courtesy of Wenner-Gren Foundation.)

described in detail, but its general features such as a long, low-vaulted skull with massive brow ridges and an angulated occipital region resemble Asian *H. erectus*. Actually, the brow ridges are more robust than in other *H. erectus* (probably a racial variation), and in this feature Hominid 9 resembles some of the late Pleistocene East and South African crania. Population continuity through time seems to be indicated in this area, too.

A skull from Bed IV at Olduvai (Hominid 12), found in 1962, has not yet been described fully (Leakey and Leakey, 1964). It is fragmentary and very thick. According to Holloway (personal communication), this skull has a low volume. The volume of the Hominid 9 skull is about 1,000 cm^3, and such a range might indicate either intrapopulation variation (seven skulls at Choukoutien yielded capacities between 850 and 1,300 cm^3) or possibly population difference. An alternative and less likely hypothesis is that these represent different species.

Judging from the associated faunal remains of large mammals in upper Bed II (Leakey, 1965), the earliest East African *H. erectus* were very competent hunters of big game.

The relationship between the Olduvai *H. erectus* and *A. habilis* has been a subject of some debate. Dr. Louis Leakey believes that they are sampled from different lineages and that Hominid 16 from lower Bed II represents a "proto-*H. erectus.*" Tobias has also suggested the possibility that they may not represent directly ancestral–descendant populations. This depends on how one interprets "directly." It is possible that there were considerable population migrations during the early and middle Pleistocene in Africa. *A. habilis* and Hominid 9 could represent different "*subspecific*" lineages within a single evolving species continuum, yet still be part of the same lineage at a species level.

Dr. Mary Leakey (1967) has studied the artifacts in Bed II and has discerned what she believes to be two industries, a "developed Oldowan" that she associates tentatively with *A. habilis* and an early Acheulean industry of tools manufactured, she believes, by *H. erectus*. The main difference between the two industries is the presence in the latter of large hand axes. The early Acheulean appears quite suddenly in East Africa, and Dr. Leakey believes that this is due to a migration into the area from elsewhere.

In 1932, an expedition led by Dr. Louis Leakey (1933) to Kanam in Kenya discovered a hominid mandible in which two premolars were preserved. When first found the mandible was thought to come from early Pleistocene deposits, although it seems rather more likely now that it is much younger. Leakey believed that the Kanam mandible exhibited certain modern morphological features, but in a recent restudy Tobias (1960) suggests that the supposed features of resemblance to *H. sapiens* (such as the development of a chin) are pathological, and he believes that the affinities of the Kanam jaw lie more with forms such as Rabat—intermediates between *H. erectus* and archaic man.

No skeletal remains of *H. erectus* have been found in South Africa, unless *"Telanthropus"* from Swartkrans represents that species.

Stone artifacts appear to be absent from European early Pleistocene deposits. The Vallonet Cave in Southeastern France may be of latest early Pleistocene age—perhaps 1 million years old—and contains a few stone tools and some fractured and flaked animal bones (Coon, 1962). Another early site, at Přezletice near Prague in Czechoslovakia, has yielded stone tools and a chip of hominid tooth (Fejfar, 1969). However, hominid occupation of Europe was probably very sparse throughout the middle Pleistocene, at least during the earlier times. Industries with hand axes appeared some 300,000 or more years ago, whereas before that the stone tools were of general Oldowan form; these types persisted in Europe well into the middle Pleistocene (Oakley, 1961).

Acheulean living sites have been excavated recently by F. Clark Howell (1965) at Torralba and Ambrona some 200 kilometers north of Madrid in Spain. Many tools have been found together with the remains of large mammals, including elephants. The hominids, probably *H. erectus*, apparently utilized fire to drive groups of large mammals into swamps where they could be easily killed and butchered. Fire was probably also used for cooking. Such activities imply a highly developed social organization, perhaps involving cooperation among several bands of hominids.

A site of similar age preserving about twenty successive living floors has recently been discovered at Terra Amata in Nice by Dr. Henri de Lumley. Evidently, hominids were constructing quite substantial dwellings at this time, because oval arrangements of stone blocks together with the remains of postholes point to the construction of temporary shelters made of poles covered by branches or hides. These shelters were about 30 by 15 feet and contained hearths and areas for cutting up food, possibly even stones to be used as seats. These hominids, although living close to the sea, were hunters with a taste for venison, elephant, and wild boar. Terra Amata is probably a site that was revisited regularly, perhaps even by the same hominid band and their descendants. The band size implied by these habitation structures is around fifteen or twenty, possibly composed of two or three nuclear families.

The oldest hominid remains from Europe, other than the tooth chip from Přezletice, come from Mauer, near Heidelberg in Germany, and from Vértesszöllös, west of Budapest in Hungary. The Mauer mandible was found in 1907. It appears to date from the Mindel glaciation, and its age is somewhere between 250,000 and 450,000 years. The jaw bone is robust and has a very broad ascending ramus. There is no chin. The teeth are small, in the lower part of the size range for *H. erectus*, but there seems to be no good reason for believing that the Mauer mandible represents any other lineage. It does foreshadow in many ways later European hominids (Howell, 1960).

The Vértesszöllös site preserves several hearths, some Oldowan-type tools, and several pieces of hominid. In 1965, remains of four milk teeth and an adult occipital bone were discovered. These specimens have been studied by Dr. Andor Thoma (1966). Their age is approximately the same as that of the Mauer jaw; that is, they are probably younger than *H. erectus* from Choukoutien, Indonesia, and Olduvai (Oakley, 1966). Thoma has concluded that the teeth show resemblances to *H. erectus* from China and differ in a number of ways from those of other hominids.

The occipital bone is much more difficult to interpret. Thoma has named it *Homo* (*erectus* seu *sapiens*), apparently because it has the morphology of *H. sapiens* although it comes from deposits of middle Pleistocene age. He has estimated the cranial capacity of the skull from which the occipital came at over 1,400 cm^3, which is larger than that of any known *H. erectus*. Although this value is based on very fragmentary evidence, the bone clearly does come from a large skull, within the range of *H. sapiens*. This may simply be because it is younger than other *H. erectus* or because it comes from a population with greater mean brain volume. However, it is the morphology that is so surprising, for the occipital bone does not resemble that of *H. erectus*, or even archaic man, but instead that of earliest modern man. Such forms are dated elsewhere as no older than 100,000 years. Possibly this implies that certain middle Pleistocene populations in Europe (and elsewhere?) contained individuals whose skull morphology, at least at the back of the cranium, was very advanced. Why should this be? Skull shape in the occipital region is determined by brain size and shape and by the size of the neck muscles attached to the outer aspect of the bone. Factors affecting brain shape are not understood at present, some large-brained hominids having brains short from front to back and rounded in the occipital region, and others equally large-brained having long brains with projecting posterior areas. These differences are reflected in the shape of the bones covering the brain. At present it is impossible to say whether the Vértesszöllös fragment represents a population ancestral specifically to modern man or whether it merely represents one end of the normal range of variation present in middle Pleistocene hominids. Only more material will enable us to answer this question.

Other middle Pleistocene hominids, younger than Mauer and Vértesszöllös, are known from Europe. These are best considered in the next chapter.

Later Pleistocene Hominids 8

The major gains made by the middle Pleistocene hominids over those of earlier times were a larger brain and a postcranial skeleton fully adapted to upright walking. These are changes that were continuations of "trends" (viewed retrospectively) commencing in the Pliocene. The major alterations affecting hominids during the last 200,000 to 300,000 years were restricted almost entirely to the head. Principally, the brain became larger, reaching average population values of 1,200 to 1,400 cm³ (approximately 300 cm³ larger than *H. erectus*). In fact, this characteristic alone is the major taxonomic factor considered in assigning fossils to *H. sapiens*. Late Pleistocene hominids also had somewhat smaller teeth, jaws, and faces, less robust brow ridges, and more rounded occipital regions than did *H. erectus* (Le Gros Clark, 1964).

Increasing brain size probably reflects mainly changes in the internal organization of the brain and alterations in the morphology of its basic units, the neurons and glial cells. These in turn are intimately related to increasingly varied individual and group behavior. Not only did hunting become more efficient, but the complexity of all aspects of man's cultural behavior expanded enormously.

Modern men appear in the fossil record some 50,000 years ago, and their ancestors can be traced back at least as far again. These men were undoubtedly hunter–gatherers; by about 40,000 years ago, modern man was very widely distributed throughout the Old World. Somewhat more than 10,000 years ago, certain Near Eastern populations shifted their subsistence patterns from hunting to a more settled farming way of life. If we assume that hunting began at least 3 million years ago, then well over 99 per cent of man's history has been spent hunting and gathering.

Modern man differs from other hominids primarily in skull morphology. The skull is rather shorter and higher than in *H. erectus* and archaic man, probably because of differences in brain shape, although alterations in facial size, and proportions and orientation, are also important. The functional significance of these features is unknown (Le Gros Clark, 1964).

Continuity between middle and early late Pleistocene hominids

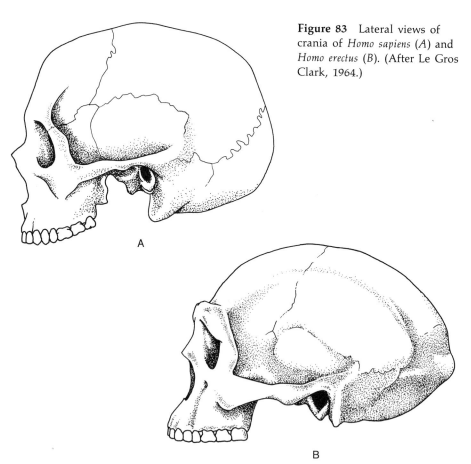

Figure 83 Lateral views of crania of *Homo sapiens* (*A*) and *Homo erectus* (*B*). (After Le Gros Clark, 1964.)

A

B

THE ASCENT OF MAN

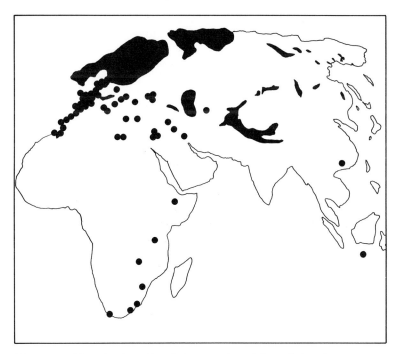

Figure 84 Distribution of archaic-man sites. Most of the sites are concentrated in Europe, although most of the hominids would have been living outside Europe. (After Howell, 1965.)

is apparent in most areas studied. That is, in any given region, *H. erectus* populations were broadly speaking ancestral to those of archaic man. The changes involved almost entirely differences in brain size and skull, face, and dental proportions. Why did these changes occur? Some workers have suggested that the evolution of populations from the *H. erectus* to the *H. sapiens* level in Indonesia, for example, was due to migration into the area by more "advanced" hominids (Coon, 1962). This still leaves a need to explain the origins of these hypothetical migrants, presumably also derived from *erectus*-like ancestors! Changes of approximately similar type over a wide area might be due to broadly similar selective pressures in all parts of the species distribution. In fact, the change from *erectus* to *sapiens* level throughout the late Pleistocene was probably the result of varying mixtures of both local evolution and migration.

The earliest archaic-man populations differed from both *H. erectus* and modern man. They had brains generally within the range of size variation of modern man and therefore larger than those of *H. erectus*, but their skulls differed morphologically from those of modern man. They had longer, lower, broader skulls, with more sharply curved occipitals, bigger brow ridges, and larger faces and teeth. The first of these archaic-man groups to be discovered and recognized, beginning in the 1850s, were the neandertals of Europe. Much later, roughly similar types were discovered in other parts of the world: Indonesia, China, and Africa. They were described variously as "Neanderthals," "Neandertaloids," "tropical Neandertalers," and so forth, as though they were in some way migrants from Europe. Had the African forms been found first, the perspective would have been different. Essentially, of course, these groups are sampled from different subspecies within a single widely ranging species, and no one of them is any more archetypal than the others. They share similar primitive features, listed above, and are therefore best described collectively as "primitive H. sapiens" or simply "archaic man."

This matter of perspective is a particularly important one, particularly in reference to the neandertals. Many more remains of archaic man have been found in Europe than in other parts of the world because there were many more anthropologists and archeologists working there. Sites were also more easily located. However, as an estimate, something like five to ten times as many hominids would have been living in habitable parts of Africa (assuming constant population densities of around twenty to forty individuals per 100 square miles) as in Europe during nonglacial periods. During glacial times perhaps no more than 5 per cent of the world's hominid populations would have been European, and for this reason any theories suggesting that Europe was a center of hominid evolution are quite likely to be incorrect.

It is probable that not all H. erectus populations evolved into archaic man, and it is also likely that only some archaic-man populations were ancestral to modern man. Let us now review the evidence from various regions.

SOUTHEAST ASIA

Between 1931 and 1941 the remains of eleven skulls—lacking faces—and two incomplete tibiae were found at Ngandong near the River Solo in Java in deposits of late Pleistocene age. The Ngandong fauna is anything from 50,000 to 250,000 years old; unfortunately, more exact dating is not possible. These hominids have been given a variety of taxonomic labels, and they are an intriguing group, for they were a late population showing many primitive features (Weidenreich, 1951).

They are clearly related to earlier Indonesian populations of H. erectus, but they had larger brains with volumes averaging around 1,100 to 1,200 cm^3. Like H. erectus specimens, the skulls still have large brow ridges and angulated occipitals and are long and low, although the breadth has increased relative to that of earlier populations. In all probability, the Solo population was descended from H. erectus of the Trinil and Djetis faunal assemblages, and it is quite possible that evolution occurred in relative isolation without much inflow of genes from other areas.

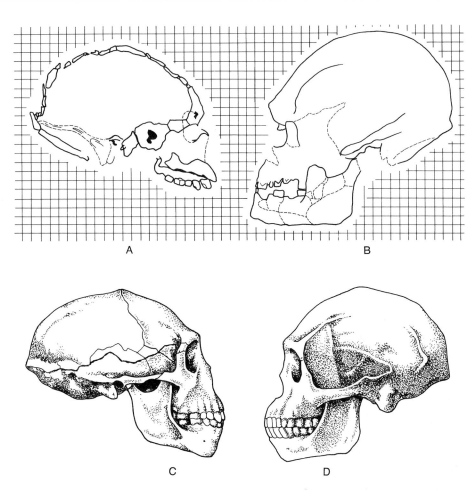

Figure 85 Comparison of skulls of archaic Australian aborigines (*A* and *B*) with *Homo erectus* (*C*) and Solo man (*D*), showing their basic similarities (C and D, faces reconstructed). (After Macintosh, 1967.)

The predominant change was one of increasing brain size occurring presumably because selection pressures favored more complex individual and group behavior. Such pressures were probably acting on all hominids throughout the Pleistocene, and lineages in widely separated areas could, and no doubt did, increase in brain size without necessarily being in contact with other more "advanced" groups.

There has been a considerable amount of discussion on the taxonomic status of the Solo population, some workers arguing that they were *H. erectus*, others believing that *H. sapiens* is a more appropriate taxonomic label (Coon, 1962). Much depends upon the biological relationships between them and other late Pleistocene hominids. If they were a lineage geographically isolated from all others for a sufficient period of time to have achieved genetic isolation of sufficient magnitude, then they could be regarded as a different species from other late Pleistocene populations. However, if there is evidence indicating that the Solo men were merely a subspecies within a highly polytypic species, then they must be considered *Homo sapiens*—whatever their morphological features. Such evidence as we have does seem on balance to favor the second interpretation, for there are indications that the Solo hominids were sampled from a lineage that evolved ultimately into populations resembling the Australian Aborigines.

In 1959, a youth's skull was found in a cave at Niah in Borneo (Brothwell, 1961). The deposits in which it had lain have been dated radiometrically at around 40,000 years old. The skull is clearly modern, and, according to Don Brothwell of the British Museum of Natural History, it resembles that of the now-extinct Tasmanians.

Man first reached Australia at least 30,000 years ago, and it is quite likely that it took several waves of migrants to populate the continent. Professor Macintosh (1967) of Sydney University has been restudying the available remains of early Aborigines during the past decade and has concluded that they indicate links with the more ancient Solo population of Java. Living Aborigines still show so-called primitive features, at least in the skull; for example, they have prominent brow ridges and often rather flattened frontal regions. However, as previously emphasized, these characteristics have never been analyzed functionally or developmentally, and it is quite possible that there are very good reasons for their retention in modern Aborigines. There is no evidence to indicate that these Australians are "primitive" in other features.

In summation, we can state that there is a fair probability that the populations of Java and, by inference, other parts of Southeast Asia form an ancestral–descendant sequence from middle Pleistocene times until the present. However, it is unlikely that these populations were evolving in total isolation, and we can assume that there was gene flow from the more northern parts of Asia throughout this time.

EUROPE

We have already discussed the history of the hominids in Europe up to the middle Pleistocene. The earlier populations appear to have been related to those in other parts of the world, and they can be regarded as *H. erectus* (although not "typical" *H. erectus* in the sense that they were identical with those Asian *H. erectus* populations discovered first). It is also quite likely that the hominids from Mauer and Vértesszöllös are in fact considerably younger than those from Java and Choukoutien, and this would account for their more "advanced" morphology.

By the end of the Mindel-Riss interglacial period, populations

Figure 86 Tentative times and durations of European glaciations together with times of origin of glaciations in other areas. (Pilbeam.)

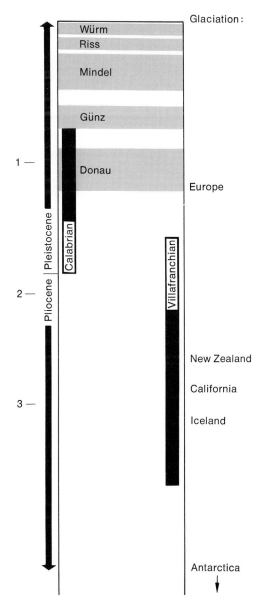

of archaic man were present in Europe, as is evidenced by specimens from Steinheim in Germany and Swanscombe in England (Le Gros Clark, 1964; Howell, 1960). Both are around 150,000 to 200,000 years old. The Steinheim specimen is the more complete; it has moderate brow ridges, a rather low brain case, and a more rounded occipital region than *H. erectus*. Brain volume was around 1,200 cm³. The morphology of the skull is intermediate between that of *H. erectus* and archaic man. The Swanscombe specimen consists of an occipital and two parietal bones. When first discovered it was thought to be similar to modern man, and an analysis

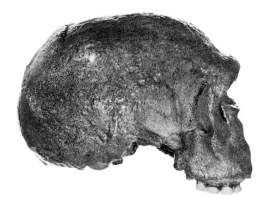

Figure 87 Side view of archaic *Homo sapiens* skull from Steinheim (cast). (Pilbeam.)

of individual measurements one by one does suggest this. However, when the measurements are combined in a multivariate statistical analysis, a different picture emerges (Weiner and Campbell, 1964). The Swanscombe and Steinheim specimens are very similar to each other and also resemble, in all measurements that can be taken, the later European hominids known as the neandertals. There is no reason to assume that these late middle Pleistocene forms, together with a mandible from Montmaurin in France, were not ancestral to the neandertals.

A series of hominids has come from deposits of the succeeding interglacial period, the Riss-Würm. These finds are between 70,000 and 100,000 years old (Howell, 1957). The most important are those from Ehringsdorf in Germany, Saccopastore in Italy, and Fonté-chevade in France. The Ehringsdorf and Saccopastore specimens are relatively complete crania, although the former lacks a face. Both resemble the Swanscombe and Steinheim specimens in some respects and later neandertals in others. The Fontéchevade finds are a little enigmatic. One specimen consists of the top of a brain case with parts of the frontal bones; the second is part of a frontal bone. The skull cap resembles the neandertals', although there is some suggestion that brow ridges were only poorly developed (Weiner and Campbell, 1964). The frontal fragment shows little or no brow ridge, a condition atypical for European hominids of this period. It has been suggested that the Fontéchevade specimen belonged to an ancient population of modern man; however, more probably it represents part of the range of variation of archaic men living in Europe and adjacent areas at that time.

The problem in dealing with such small samples is that population movements into as well as within Europe are undetectable, at least by using such small skeletal samples. The evidence does not support the view that there were two distinct lineages through-out the middle and late Pleistocene, one leading to neandertal man and the other to modern man. As noted above, there is no reason to believe that more than one subspecific lineage is being sampled in Europe and every reason to think that this lineage was ancestral to the "classic" neandertals.

Figure 88 Side view of Würm age neandertal from Monte Circeo in Italy. (Courtesy of I. Tattersall.)

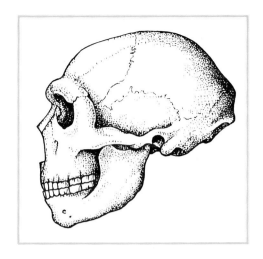

Figure 89 Mousterian stone tools (below) and Upper Paleolithic tools (above). All the tools are from Europe. (Pilbeam.)

Figure 90 Side view of Würm neandertal skull from La Ferrassie in France. (Courtesy of I. Tattersall.)

Neandertals have been divided into two general groups (Howell, 1957): those found in Western Europe during the first part of the last glaciation (approximate ages of between 75,000 and 45,000 years) forming a rather homogeneous group morphologically; and all the others that have also been called neandertals! The Western European neandertals were relatively specialized compared with other late Pleistocene hominids. They had long low skulls, with prominent occiputs and well-developed brow ridges atop jutting faces. They are associated with a complex of stone tool industries known as the Mousterian (Oakley, 1961). They were fully upright bipeds, although of stocky build; their postcranial skeletons exhibit certain peculiarities, as yet poorly understood, differentiating them from other groups of *H. sapiens.*

For some time it was believed that these neandertals were brutish and subhuman creatures, barely capable of walking erect. In fact, nothing could be further from the truth (Straus and Cave, 1957). They made very complex stone tools, they hunted large mammals, including huge cave bear, with great skill, they buried their dead with ceremony, and they colonized Western Europe in the bitter cold of the last glaciation. Assuredly, they were fully sapient human beings. Their importance has been greatly overstressed in human evolutionary studies for they have been studied in detail mainly because they are European, as at one time were most paleoanthropologists.

Some 40,000 years ago in Europe there was a break in tool-making traditions; new types of industries (so-called Upper Paleolithic) appeared, often quite abruptly, and it seems very likely that

Figure 91 (opposite)   Old reconstruction of European neandertal man (*A*) compared with modern man (*B*). Earlier workers believed neandertals to be brutish creatures, incompletely bipedal, showing many apelike features. This view is known now to be incorrect. Postcranially, neandertals were bipeds as efficient as any modern populations. (After Straus and Cave, 1957.)

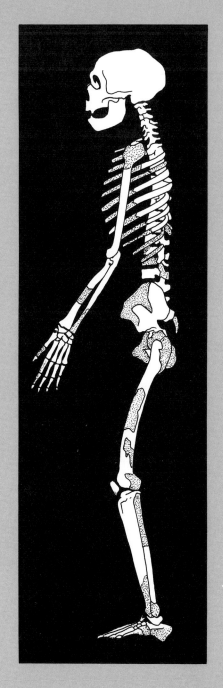

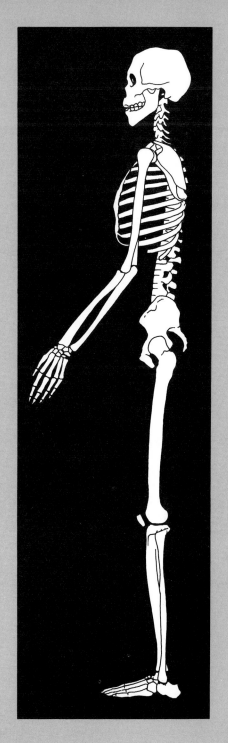

A

B

at least in Western and parts of Central Europe these new types were brought by migrants from Eastern Europe, North Africa, and Western Asia, migrants who were fully evolved modern men (Oakley, 1961).

The fate of the neandertals is unknown. Some probably died out, and others were no doubt absorbed by the new populations. It is most unlikely that they were killed by "superior" types. The phenomena of population movement, expansion, decline, absorption, and so forth were occurring all the time. Because of the accidents of discovery, we have a particularly good example of them in Western Europe.

Further to the east, there is some evidence for populations in the time period between 30,000 and 40,000 years ago that were either transitional groups or hybrids between the neandertals and modern man. Between 1899 and 1905 a series of hominids was recovered from a rock shelter at Krapina in Yugoslavia (Howell, 1957). There has been some disagreement about the dating of the Krapina remains, but it now seems likely that they belong to a warm period during the last glaciation and are approximately 30,000 to 40,000 years old (Schaefer, 1959). The remains represent more than a dozen adults and children and consist of skulls and post-cranial bones. The skulls are intermediate morphologically between the "classic" Western European neandertals and modern man, and the postcranials also show a mixture of features, although in general they resemble modern man closely. In 1965, a maxillary fragment representing an apparently similar intermediate type was found in deposits of approximately the same age at Kulna in Czechoslovakia (Jelinek, 1966).

How can these remains be interpreted? It is possible that they represent hybrids between neandertals and modern types. Alternatively, they may form part of a morphological continuum moving from neandertals in the west to more fully modern types in Eastern Europe and Western Asia. This latter assumption is perhaps the more probable, for although the evidence is extremely fragmentary it does seem likely that during the last 100,000 years or so the populations of Europe, North and East Africa, and contiguous parts of Western Asia formed a mosaic of types ranging morphologically between the specialized neandertals and forms differing very little from modern man. The further east and south we trace hominid populations at this time the more modern looking they become.

AFRICA AND WESTERN ASIA

Perhaps the most interesting changes in the middle and later Pleistocene occurred in Africa and adjacent areas of Asia, for it is here that we find traces of the very earliest modern men, *Homo sapiens sapiens*. Populations in North, East, and South Africa in the middle

Pleistocene show affinities with earlier *H. erectus* populations in these areas. In North Africa, mandibles from Temara in Morocco, from Mughharet-el-Aliya in Tangiers, and from Haua Fteah in Libya indicate that archaic-man populations lived here until 40,000 years ago (Howell, 1960). These populations may well have been descendants of the earlier groups represented by, for example, the Ternifine jaws. They have been described as "neandertaloid"— neandertal-like—but they were not neandertals. They show many resemblances to late Pleistocene specimens from East and South Africa, suggesting the existence of a whole complex of related groups of hominids spread throughout that continent.

Two relatively late skulls are known from North Africa, from Jebel Irhoud in Morocco. These show similarities to the Western European neandertals (Arambourg, 1963b). Further to the south several specimens of such archaic men have been located. Sites at Singa in the Sudan, Eyasi in Tanzania, Broken Hill in Zambia, and Saldanha in South Africa have yielded skulls or skull fragments, and another cave at Makapansgat (of *Australopithecus* fame) has provided us with a mandible (Tobias, 1961, 1962). This material demonstrates the existence in South and East Africa of an archaic type of man characterized by massive brow ridges, low vault, and projecting occipital region. In some features these specimens re-

Figure 92 *Homo erectus* (above) from the middle Pleistocene of East Africa compared with archaic man (below) from the late Pleistocene of Southern Africa, showing the basic similarities of earlier and later members of *Homo* in sub-Sahara Africa (casts). (Courtesy of Wenner-Gren Foundation.)

semble European neandertals, in others North African and Western Asian "neandertaloids"; certain features recall the Solo skulls and some (mostly postcranial characters) point to affinities with modern man. This population (or populations) was in existence at least as late as 35,000 years ago. Almost certainly these late Pleistocene hominids were the larger-brained descendants of *Homo erectus* populations that lived earlier in the same areas. For example, the similarities between Hominid 9 from Olduvai Bed II and the Broken Hill skull are rather marked.

Thus we can infer a general trend, throughout the Old World, toward the development from *H. erectus* populations of groups of archaic man classifiable as *H. sapiens.* At any one time level there would probably have been only one species, although some populations would have expanded, some would have hybridized, and some others would have become extinct. Yet we can observe continuity both through time and at any given time level.

What happened to these archaic-man populations? Some were replaced like the Western European neandertals. Others—those in Eastern Europe, China, and Southeast Asia—seem to have evolved into modern types by a combination of phyletic change and hybridization. But for hybridization to result in "modernization" at least one of the original populations would have to have been more modern in aspect. Whence came these advanced men?

Between 1929 and 1934 Professor Dorothy Garrod recovered a series of hominids from two caves, Mugharet et-Tabūn and Mugharet es-Skhūl, on Mount Carmel, south of Haifa in Israel (Howell, 1957). The dating of the caves has been debated for some time, but it now seems probable that they are between 40,000 and 60,000 years old; it has been suggested that Tabūn may be older than Skhūl by as much as 10,000 years (Higgs, 1961). Both sets of remains are associated with a type of Mousterian industry. Two individuals have been found at Tabūn, both clearly representing a population of neandertal-like hominids. However, they are less specialized—more modern in appearance—than the Western European forms and have affinities with other archaic types to the north and south as well as with a sample of (seven) hominids from Shanidar in Iraq (Solecki, 1963). The Shanidar remains range in age from around 45,000 to some 80,000 years and are also associated with a Mousterian culture.

The hominids from the Skhūl cave together with a maxillary fragment from Ksar 'Akil in Lebanon and a series of specimens from Djebel Kafzeh in Israel are of considerable interest in our search for the ancestors of modern man. All are associated with Mousterian industries, all appear to date from around 45,000 to 55,000 years ago, and all are rather modern in aspect. The ten specimens from Skhūl, adults as well as immature individuals, include specimens that are as modern as some of the earliest undoubted *Homo sapiens sapiens* from Europe, even though the Skhūl hominids are geologically older. Skulls are relatively short from front to back and well vaulted, with moderate brow ridges and

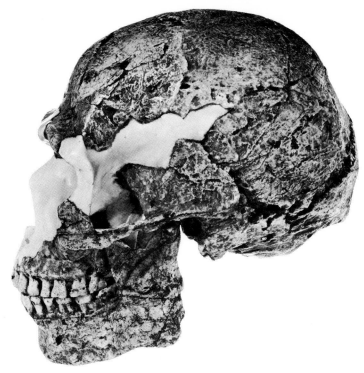

Figure 93 Skull of *Homo sapiens sapiens* from Skhūl, Israel, one of the earliest known examples of modern man (cast). (Courtesy of Wenner-Gren Foundation.)

rounded occipital regions. The facial skeleton appears to be tucked under the brain case more than in neandertals. These changes, as noted before, are the basic ones that must be accounted for in any discussion of the origins of modern man; as yet, no plausible explanations have been produced.

Apparently, the shift from Mousterian to the Upper Paleolithic type industries typically associated with modern man in Europe, North Africa, and Asia occurred after modern man had appeared. According to most workers, the shift in tool traditions indicates "new and more economical ways of utilizing raw material . . . [by] the consistent manufacture of implements by punch-blade technique and a marked increase in frequencies of composite tools." The preceding quote is from Professor Sally Binford (1968), who has suggested that the origin and eventually widespread dispersal of modern man was due to a shift, in certain populations, in social organization. She notes further that there appears to have been a change in hunting patterns toward the end of the Mousterian, at least in Western Asia, from generalized big-game hunting to the successful pursuit of large herd animals such as wild cattle (*Bos primigenius*). She believes that such a change in hunting methods

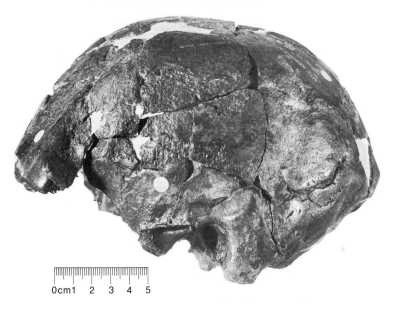

Figure 94   Hominid skull (Omo II) from late Pleistocene deposits at Omo, Ethiopia. (Courtesy of M. H. Day.)

would require changes in social organization permitting the formation of much larger social groupings than were previously necessary:

> We may therefore envision that in the change to the systematic exploitation of migratory herd mammals there was a concomitant change in human mating patterns to a broader kind of exogamous linkage between formerly self-sufficient small bands. (Binford, 1968:714)

The hunting of large herd animals is characteristic of the European Upper Paleolithic. Binford believes that the shift from Mousterian to later industries was rapid and the result of the earlier social and economic changes. Did these admittedly hypothetical social changes cause the morphological change to modern man, or did they simply occur, by accident as it were, in populations that had more modern morphologies than others? This is a difficult question to answer, although there is some evidence that indicates the presence of modern-looking (at least relatively speaking) populations at least as far back as 100,000 years ago.

In 1967, Richard Leakey (1969) recovered remains of three hominids from the lower levels of the Kibish formation in the Omo region of Ethiopia. The deposits are perhaps as much as 90,000 or 100,000 years old, and the remains, although showing some affinities with African archaic man, are surprisingly modern looking. Michael Day, who has published a preliminary description of these specimens (Day, 1969b), summarizes as follows:

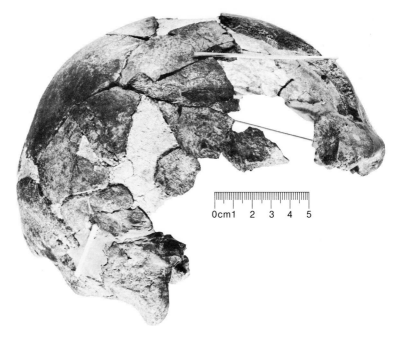

Figure 95 Hominid skull (Omo I) from late Pleistocene deposits at Omo, Ethiopia. (Courtesy of M. H. Day.)

The more complete calvaria, Omo II, has many features both in its general configuration and in its detailed anatomy, which is similar to the Solo skulls and, to a lesser extent, the Broken Hill skull, the Vértesszöllös occipital, the Kanjera skulls, and even indeed *Homo erectus.* On the other hand, the Omo I skull, which is contemporaneous with Omo II, is more modern in its general form and can be reasonably compared with . . . [the] Skhūl skulls. (Day, 1969a:1137)

The Kanjera specimens mentioned in the quote were recovered in Kenya by Dr. Louis Leakey in 1932 (Leakey, 1933). Their age has been vigorously debated since their discovery, but it now seems more probable than not that they are at least as ancient as the Omo material (Tobias, 1962). Although they resemble other early hominid crania in the occipital region in that they lack the rounded contour typical of the cranium of modern man, unlike neandertaloids they do not possess prominent brow ridges. It would appear then that populations containing individuals who were at least partially modern in appearance existed at least 100,000 years ago in East Africa.

Brothwell (1963) has proposed that the area of East Africa, Arabia, Western Asia, and India was the "cradle" of *Homo sapiens sapiens,* although such a hypothesis is based on almost no concrete data. However, it is probable that during the period between 100,000 and 50,000 years ago many populations within this area did evolve in the direction of modern man, the changes being confined mainly to the skull, face, jaws, and teeth. Shifts in social organization, and also in social behavior, probably occurred during this time, perhaps associated with an increasing frequency of hunting large herd mammals. The time between 50,000 and 30,000 years ago saw the spread of modern man out of his hypothetical "Garden of Eden" until, through a process of swamping and replacing older and more archaic subspecies of *H. sapiens*, he inherited the earth.

Conclusion 9

In this final chapter we shall review briefly what is known of the course of human evolution and try to discern some of the "growth areas" where future research is likely to be profitable.

In the main, the early primates were arboreal animals living in tropical and subtropical rainforests, feeding on a variety of vegetable foods although supplementing their diet with animal protein. Before the emergence of the higher primates, certain groups of prosimians were almost certainly living in cohesive social groups that were more or less permanent year-long aggregations of adults and their offspring. These primates were diurnal, agile creatures, with stereoscopic color vision and highly developed manipulative skills.

During the Oligocene and Miocene, man's ancestors probably resembled closely in overall evolutionary grade the living monkeys of the New and Old Worlds. The most marked advance over their prosimian ancestors would have been in behavior: these ancestors of ours were large-brained—for their time the largest-brained animals living—and the learned component of their behavior was much greater than in any other mammal. This growth in behavioral complexity was probably linked to their evolution as social mammals. To maintain social cohesion within the group, communication between individuals became of prime importance. Most of the information conveyed involved the internal emotional or motivational state of the animal; it was carried through a number of channels, involving vocalizations, postures, and gestures, as well as the older olfactory signals. Each animal not only produced a stream of such information but also continually monitored the behavior of others in the group; this monitoring enabled the status-conscious (or "dominance"-conscious) primates to maintain the stability of the troop, each animal learning and reacting to the behavior of the others while producing appropriate responses in every context.

Behaviorally, at least in terms of social behavior, man's immediate ancestors were probably not unlike chimpanzees. And ecologically, too, prehominids are likely to have been forest animals capable, like the chimpanzee, of life at the forest fringe and in open woodland. Indirect evidence both from studies of living primates and from the little that is known of *Australopithecus* postcranial anatomy suggests that the Miocene prehominids were arm-swinging forms, fully capable of suspending themselves below supports while feeding and moving in the trees. On the ground, they might have been knuckle walkers, and they may well also have moved bipedally.

Late in the Miocene, certain dramatic behavioral changes occurred, resulting in the emergence of Hominidae. The dentition became drastically altered as a response to a new diet and a totally novel pattern of feeding. At present, the nature of the shifts is somewhat obscure, but they appear to have been related to a shift toward ground feeding—at first within the forest and at its fringe, but later increasingly in more open country. The type of food eaten also changed, from a predominantly frugivorous diet to one composed of tougher, smaller vegetable items; it is also quite possible that the new diet contained significantly greater quantities of meat.

Late Miocene African and Asian species of the genus *Ramapithecus* already possessed the most important hominid dental changes and were presumably becoming adapted to a new ecological niche, a new diet, and a new pattern of living. Unfortunately, no postcranial bones of *Ramapithecus* species are known, yet it can be assumed that these were creatures that did not possess the bipedal adaptations of later hominids but rather were still basically arm-swinging forms.

Little is known of the Hominidae of between 11 million and 4 million years ago, but it seems clear that during this time our ancestors moved from forests into the woodlands and spreading savannas of Africa, and possibly Asia, too. Probably the two most important adaptations that then developed were (1) a trend toward further meat eating requiring hunting and involving food sharing, division of labor between sexes, more refined communications systems, and so forth, and (2) continuing improvement in the vertebral column, pelvis, and lower limbs associated with the acquisition of habitually erect bipedal locomotor pattern.

Around 4 million years ago at least two distinct species lineages were in existence: one, relatively large-toothed (*Australopithecus robustus*), the other smaller-toothed (*A. africanus*). These are known from both East and South Africa throughout the Pliocene and early Pleistocene. When these two lineages diverged is unknown. *A. robustus*, judging from its dental and cranial adaptations, was probably a more herbivorous form than *A. africanus*; apparently, it became extinct during the Pleistocene. The smaller species, *A. africanus* (weighing around 50 pounds), was almost certainly a tool maker and a hunter; it was fully bipedal, although possessing forelimbs suggesting that it was still capable of suspensory locomotion at least occasionally. There is also evidence indicating that the hominid brain had expanded during the Pliocene.

If we assume that *Ramapithecus* species were pygmy chimpanzee size (around 50 pounds mean body weight) and if we further assume that their brain size was also similar to that of the pygmy chimpanzee (some 300 cm^3), then it is apparent that the Pliocene saw a considerable expansion in brain size, up to at least 440 cm^3, with little or no change in body weight. The change in brain size was probably correlated with internal reorganization, and the main factors involved in the reorganization are likely to have been the development of tool-making abilities and the evolution of language. Further expansion of the brain continued during the latest Pliocene, again with little or no change in body size. The "descendant" of *A. africanus, A. habilis,* had an average brain volume of around 600 cm^3. After the early Pleistocene, changes in hominid morphology and behavior became accelerated, and it is probably best to treat *A. habilis* as the final part of the long initial stages of hominid evolution. It now seems that during these earlier phases the basic hominid behavioral and morphological features were established and that the terminal stages of human evolution saw the elaboration and refinement of these characteristics.

Briefly, the most important features seem to have been the following: the initial dietary shifts and the correlated dental and cranial changes; the development of hunting, involving cooperation among males, and of food sharing; the formation of permanent home bases; the evolution of pair bonding, resulting in the addition of a male to the mother–offspring unit; the slower maturation of the infant, requiring longer dependence on parents—particularly the mother —and allowing greater behavioral flexibility; the development of incest taboos and exogamic mating problems; the evolution of tool making, probably correlated—at least in part—with habitual bipedal locomotion; and the development of language. The last is probably the most important feature of all, because it resulted in the appearance of "cultural" behavior, made the environment even more complex as it became increasingly subdivided and labeled (arbitrarily), and acted therefore as a catalyst for still more brain–

behavior evolution. If this was the most important change, still others were of almost equal importance. One major behavioral shift was that away from aggressive, individualistic, dominance-oriented behaviors to more cooperative patterns.

The main thrust of hominid evolution after the early Pleistocene seems to have involved the refinement of these features. *Homo erectus* and *Homo sapiens,* together with their associated artifacts, followed the path of increasingly complex cultural behavior, more effective hunting (more and bigger game), and more diverse and better-made tools. They were also larger creatures than the earlier hominids, probably an adaptation to changes in hunting behavior and to increasing home-range size. Postcranially, the hominids after *A. habilis* appear to have been indistinguishable from modern man. Of particular interest are the alterations in the entire forelimb, from scapula to fingers. These changes probably occurred because of new methods of tool making and new types of tool using.

The final stages of human evolution up to the emergence of modern man mainly involved increases in brain size, although there is some evidence to suggest that the terminal stages of hominid phylogeny saw changes in linguistic communication—possibly at the periphery (correlated with changes in the position of the larynx and in pharyngeal morphology rather than in the brain).

Therefore, the major trends in hominid evolution seem to have been, first, a shift in ecological niche and diet, and, second—possibly flowing from these earlier changes—major behavioral reorganization involving a new type of social organization and a unique mode of communication.

Much can be gained from viewing our behavior in an evolutionary perspective, for the theory of natural selection applies as much to man's behavior as to any other human feature. But human behavior is very difficult to analyze because so much of it is learned and therefore quite variable from one culture to another.

Recently there have been a number of books published in which it is claimed that certain aspects of human behavior are, if not instinctive, at least under strong genetic control. For example, it has been stated that man is innately territorial and aggressive (or nonaggressive) and that the concentration of political power in the hands of males is due to fundamental biological differences between the sexes. I am perhaps doing some injustice to the authors' views in summing them up so briefly, yet the main thrust of their arguments is that these human behavioral characteristics are to a large extent innate and not cultural.

It is now realized that all behavior is both innate and learned. Much of the behavior of nonhuman primates involves a large learning component, the genetic influences acting mainly to channel the development of a particular behavior pattern. Some behaviors are relatively hard to acquire. Others may be easy to learn and their learning pleasurable. For example, language acquisition in man is extremely difficult to suppress, yet the end results, in terms of the type of language learned, are tremendously variable.

Certainly, men can behave aggressively; they also cooperate. Some groups may be territorial whereas others are not. Pair bonds are generally fairly stable in most cultures, yet the very fact that they are not invariably stable, that economic and social factors greatly affect their duration, should serve to demonstrate that pair bonding, like the other features here listed, is subject to a whole complex of determinants. Much more careful work needs to be done on all these characteristics before it will be possible to determine to what extent human behavior is or is not innate. This will require detailed cross-cultural studies of human behavior in a variety of contexts, psychological studies of cultures other than our own, and a clearer understanding of the course of human evolution, as well as behavioral studies of nonhuman primates. Only when these data are synthesized will it be possible to answer questions about the fundamentals of human nature.

Man evolved during millions of years as a hunter, living in small groups. To what extent is modern urban man maladapted biologically to living in large, crowded communities? Can men learn to plan their future survival? These are questions of tremendous importance, and we need answers to them as quickly as possible However, the immediate answer is more research in all areas relevant to the study of man.

Bibliography

Andrew, R. J., 1964, The displays of the primates, IN *Evolutionary and genetic biology of primates,* ed. J. Buettner-Janusch, Academic, p. 227.

Arambourg, C., 1963a, Le gisement de Ternifine, *Arch. Inst. Pal. Hum.,* 32, 1.

———, 1963b, Le gisement moustérien et l'homme du Jébel Irhoud (Maroc), *Quaternaria,* 7, 1.

———, and Biberson, P., 1956, The fossil human remains from the Paleolithic site of Sidi Abderrahman (Morocco), *Amer. J. Phys. Anthrop.,* 14, 467.

———, and Coppens, Y., 1968, Découverte d'un australopithécien nouveau dans les gisements de l'Omo, *S. Afr. J. Sci.,* 64, 58.

Binford, S. R., 1968, Early upper Pleistocene adaptations in the Levant, *Amer. Anthrop.,* 70, 707.

Bishop, W. W., and Chapman, G. R., 1970, Early Pliocene sediments and fossils from the northern Kenyan Rift Valley, *Nature,* 226, 914.

Bishop, W. W., Miller, J. A., and Fitch, F. J., 1969, New potassium–argon age determinations relevant to the Miocene fossil mammal sequence in East Africa, *Amer. J. Sci.,* 267, 669.

Bordes, F., 1968, *The Old Stone Age,* World University Library.

Brain, C. K., 1958, The Transvaal ape-man-bearing cave deposits, *Transvaal Mus. Mem.,* no. 11.

———, 1967a, The Transvaal Museum's fossil project at Swartkrans, *S. Afr. J. Sci.,* 63, 378.

———, 1967b, Hottentot food remains and their bearing on the interpretation of fossil bone assemblages, *Sci. Pap. Namib Desert Res. Station,* no. 32.

———, 1970, New find at the Swartkrans australopithecine site, *Nature,* 225, 1112.

Broom, R., 1936, A new fossil anthropoid skull from South Africa, *Nature,* 138, 486.

———, 1938, The Pleistocene anthropoid apes of South Africa, *Nature,* 142, 377.

———, and Robinson, J. T., 1949, A new type of fossil man, *Nature,* 164, 322.

Brothwell, D. R., 1961, Upper Pleistocene human skull from Niah Caves, Sarawak, *Sarawak Mus. J.,* 9, 323.

———, 1963, Where and when did man become wise? *Discovery* (Lond.), p. 10.

Butler, P. M., and Mills, J. R. E., 1959, A contribution to the odontology of *Oreopithecus, Bull. Brit. Mus. Nat. Hist.,* 4, 1.

Campbell, B. G., 1966, *Human evolution,* Aldine.

Chow, M., 1965, Mammalian fossils associated with the hominid skull cap of Lantian Shensi, *Sci. Sin.,* 14, 1037.

Clarke, R. J., Howell, F. C., and Brain, C. K., 1970, More evidence of an advanced hominid at Swartkrans, *Nature,* 225, 1219.

Cooke, H. B. S., 1963, Pleistocene mammal faunas of Africa, with particular reference to Southern Africa, IN *African ecology and human evolution,* eds. F. C. Howell and F. Boulière, Aldine, p. 65.

Coon, C. S., 1962, *The origin of races,* Knopf.

Crompton, A. W., and Hiiemäe, K., 1969, How mammalian molar teeth work, *Discovery* (Peabody Mus., Yale), 5, 23.

Crook, J. H., 1966, Gelada baboon herd structure and movement: A comparative report, *Symp. Zool. Soc. Lond.,* no. 18, 237.

———, and Gartlan, J. S., 1966, Evolution of primate societies, *Nature,* 210, 1200.

Dart, R. A., 1925, *Australopithecus africanus,* the man-ape of South Africa, *Nature,* 115, 195.

———, 1934, The dentition of *Australopithecus africanus, Folia Anat. Japon.,* 12, 207.

———, 1948, The Makapansgat protohuman *Australopithecus prometheus, Amer. J. Phys. Anthrop.,* 6, 259.

———, 1957, The osteodontokeratic culture of *Australopithecus prometheus, Transvaal Mus. Mem.,* no. 8.

———, 1962, The Makapansgat pink breccia australopithecine skull, *Amer. J. Phys. Anthrop.,* 20, 119.

Davenport, R. K., 1967, The orangutan in Sabah, *Folia Primat.,* 5, 247.

Davis, P. R., 1964, Hominid fossils from Bed I, Olduvai Gorge, Tanganyika: A tibia and fibula, *Nature,* 201, 967.

Day, M. H., 1967, Olduvai Hominid 10: A multivariate analysis, *Nature,* 215, 323.

———, 1969a, Femoral fragment of a robust australopithecine from Olduvai Gorge, Tanzania, *Nature,* 221, 230.

———, 1969b, Omo human skeletal remains, *Nature,* 222, 1135.

———, and Napier, J. R., 1964, Hominid fossils from Bed I, Olduvai Gorge, Tanganyika: Fossil foot bones, *Nature,* 201, 968.

———, and Wood, B. A., 1968, Functional affinities of the Olduvai Hominid 8 talus, *Man,* 3, 440.

———, 1969, Hominoid tali from East Africa, *Nature,* 222, 591.

Dubois, E., 1894, *Pithecanthropus erectus, eine menschenähnliche Übergangsform aus Java,* Batavia, Landesdruckerei.

———, 1924, On the principal characters of the cranium and brain, the mandible and the teeth of *Pithecanthropus erectus, Proc. Acad. Sci. Amst.,* 27, 265.

Ellefson, J. O., 1968, Territorial behavior in the common white-handed gibbon, *Hylobates lar* Linn, IN *Primates,* ed. P. C. Jay, Holt, p. 180.

Erikson, G. E., 1963, Brachiation in New World monkeys and in anthropoid apes, *Symp. Zool. Soc. Lond.,* no. 10, 135.

Evernden, J. F., and Curtis, G. H., 1965, The potassium–argon dating of late Cenozoic rocks in East Africa and Italy, *Curr. Anthrop.,* 6, 343.

Every, R. G., 1970, Sharpness of teeth in man and other primates, *Postilla* (Yale), no. 143.

Fejfar, O., 1969, Human remains from the early Pleistocene in Czechoslovakia, *Curr. Anthrop.,* 10, 170.

Freedman, L., 1963, A biometric study of *Papio cynocephalus* skulls from N. Rhodesia and Nyasaland, *J. Mammal.,* 44, 24.

Gardner, R. A., and Gardner, B. T., 1969, Teaching sign language to a chimpanzee, *Science,* 165, 664.

Goodall, J., 1965, Chimpanzees of the Gombe Stream Reserve, IN *Primate behavior,* ed. I. DeVore, Holt, p. 425.

Goodhart, C. B., 1960, The evolutionary significance of human hair pattern and skin colouring, *Adv. Sci.,* 17, 53.

Gregory, W. K., 1916, Studies on the evolution of the primates, *Bull. Amer. Mus. Nat. Hist.,* 35, 239.

Hall, K. R. L., 1965a, Behavior and ecology of the wild patas monkey, *Erythrocebus patas,* in Uganda, *J. Zool.,* 148, 15.

———, 1965b, Social organization of the Old World monkeys and apes, *Symp. Zool. Soc. Lond.,* no. 14, 265.

———, and DeVore, I., 1965, Baboon social behavior, IN *Primate behavior,* ed. I. DeVore, Holt, p. 53.

Hay, R. L., 1967, Revised stratigraphy of Olduvai Gorge, IN *Background to evolution in Africa,* eds. W. W. Bishop and J. D. Clark, U. of Chicago, p. 221.

Hayes, K. J., and Hayes, C., 1955, The cultural capacity of chimpanzee, IN *Nonhuman primates and human evolution,* Wayne State University, p. 110.

Higgs, E. S., 1961, Some Pleistocene faunas of the Mediterranean coastal areas, *Proc. Prehist. Soc.,* 27, 144.

Hockett, C. F., 1960, The origin of speech, *Sci. Amer.,* 203, 88.

Holloway, R. L., 1968, Cranial capacity and the evolution of the human brain, IN *Culture: Man's adaptive dimension,* ed. M. F. A. Montagu, Oxford, p. 170.

———, 1969, Culture: A human domain, *Curr. Anthrop.,* 10, 395.

———, 1970, Australopithecine endocast (Taung specimen, 1924): A new volume determination, *Science,* 168, 966.

Bibliography

Hooijer, D. A., 1948, Prehistoric teeth of man and the orangutan from Central Sumatra, with notes on the fossil orangutan from Java and South China, *Zool. Meded.,* 29, 175.

————, **1960,** Quaternary gibbons from the Malay Archipelago, *Zool. Verhand.,* no. 46.

————, **1962,** The middle Pleistocene fauna of Java, IN *Evolution und Hominisation,* ed. G. Kurth, Fischer, p. 108.

Howell, F. C., 1955, The age of the australopithecines of Southern Africa, *Amer. J. Phys. Anthrop.,* 13, 635.

————, **1957,** The evolutionary significance of variation and varieties of "neandertal" man, *Q. Rev. Biol.,* 32, 330.

————, **1960,** European and Northwest African middle Pleistocene hominids, *Curr. Anthrop.,* 1, 195.

————, **1965,** *Early man,* Life Nature Library, Time-Life, Inc.

————, **1968,** Omo research expedition, *Nature,* 219, 567.

————, **1969,** Remains of Hominidae from Plio-Pleistocene formations in the lower Omo Basin, Ethiopia, *Nature,* 223, 1234.

Howells, W. W., 1966, *Homo erectus, Sci. Amer.,* 215, 46.

Hrdlička, A., 1935, The Yale fossils of anthropoid apes, *Amer. J. Sci.,* 29, 34.

Hürzeler, J., 1954, Contribution à l'odontologie et à la phylogénie du genre *Pliopithecus* Gervais, *Ann. Paleont.,* 40, 5.

————, **1958,** *Oreopithecus bambolii* Gervais: A preliminary report, *Vert. Naturf. Ges. Basel,* 69, 1.

Imanishi, K., 1960, Social organization of subhuman primates in their natural habitat, *Curr. Anthrop.,* 1, 393.

Isaac, G. Ll., 1967, The stratigraphy of the Peninj group: Early Pleistocene formations west of Lake Natron, Tanzania, IN *Background to evolution in Africa,* eds. W. W. Bishop and J. D. Clark, U. of Chicago, p. 229.

————, **1969,** Studies of early culture in East Africa, *World Arch.,* 1, 1.

Itani, J., and **Suzuki, A., 1967,** The social unit of chimpanzees, *Primates,* 8, 355.

Jacob, T., 1964, A new hominid skullcap from Pleistocene Sangiran, *Anthropologia,* 4, 97.

Jelinek, J., 1966, Jaw of an intermediate type of neandertal man from Czechoslovakia, *Nature,* 212, 701.

Jolly, A., 1966a, *Lemur behavior,* U. of Chicago.

————, **1966b,** Lemur social behavior and primate intelligence, *Science,* 153, 501.

Jolly, C. J., 1967, The evolution of the baboons, IN *The baboon in medical research,* vol. 2, p. 23.

————, **1970a,** The seed eaters: A new model of hominid differentiation based on a baboon analogy, *Man,* 5, 5.

————, **1970b,** The large African monkeys as an adaptive array, IN *Old World Monkeys,* eds. J. R. Napier and P. H. Napier, Academic, p. 139.

Kohl-Larsen, L., 1943, *Auf den Spuren des Vormenschen,* Stuttgart, 2, 379.

Kohne, D. E., 1970, Evolution of higher-organism DNA, *Q. Rev. Biophys.,* 3, 327.

Kurtèn, B., 1960, On the date of Peking man, *Soc. Sci. Fenn. Comm. Biol.,* 23, no. 7.

Lancaster, J. B., 1968, Primate communication systems and the emergence of human language, IN *Primates,* ed. P. C. Jay, Holt, p. 439.

Le Gros Clark, 1947, Observations on the anatomy of the fossil Australopithecinae, *J. Anat.,* 81, 300.

————, **1959,** *The antecedents of man,* Edinburgh.

————, **1964,** *The fossil evidence for human evolution,* 2nd. ed., U. of Chicago.

————, and **Leakey, L. S. B., 1951,** The Miocene Hominoidea of East Africa, *Fossil mammals of Africa,* no. 1, Brit. Mus. Nat. Hist.

————, and **Thomas, D. P., 1951,** Associated jaws and limb bones of *Limnopithecus macinnesi, Fossil mammals of Africa,* no. 3, Brit. Mus. Nat. Hist.

Leach, E., 1967, *A runaway world?,* BBC.

Leakey, L. S. B., 1933, The status of the Kanam mandible and the Kanjera skulls, *Man,* 33, 200.

————, **1958,** Recent discoveries at Olduvai Gorge, Tanganyika; *Nature,* 181, 1099.

————, **1959,** A new fossil skull from Olduvai, *Nature,* 184, 491.

————, **1961,** New finds at Olduvai Gorge, *Nature,* 189, 649.

————, **1962,** A new lower Pliocene fossil primate from Kenya, *Ann. Mag. Nat. Hist.,* 4, 689.

————, **1965,** *Olduvai Gorge 1951–1961,* vol. 1, Cambridge U. Press.

————, **1966,** *Homo habilis, Homo erectus,* and the australopithecines, *Nature,* 209, 1279.

————, **1967,** An early Miocene member of Hominidae, *Nature,* 213, 155.

————, and **Leakey, M. D., 1964,** Recent discoveries of fossil hominids in Tanganyika: At Olduvai and near Lake Natron, *Nature,* 202, 3.

————, **Tobias, P. V.,** and **Napier, J. R.,**

1964, A new species of the genus *Homo* from Olduvai Gorge, *Nature,* 202, 7.

——, **and Whitworth, T., 1958,** Notes on the genus *Simopithecus,* with a description of a new species from Olduvai, *Coryndon Memorial Museum Occasional Papers,* No. 6.

Leakey, M. D., 1967, Preliminary survey of the cultural material from Beds I and II, Olduvai Gorge, Tanzania, IN *Background to evolution in Africa,* eds. W. W. Bishop and J. D. Clark, U. of Chicago, p. 417.

Leakey, R. E. F., 1969, Early *Homo sapiens* remains from the Omo River region of Southwest Ethiopia, *Nature,* 222, 1132.

——, **1970a,** In search of man's past at Lake Rudolf, *National Geog. Mag.,* 137, 712.

——, **1970b,** Fauna and artifacts from a new Plio-Pleistocene locality near Lake Rudolf in Kenya, *Nature,* 226, 223.

Lee, R. B., and DeVore, I., 1968, *Man the hunter,* Aldine.

Lenneberg, E. M., 1964, A biological perspective of language, IN *New directions in the study of language,* ed. E. M. Lenneberg, M. I. T., p. 65.

Lewis, G. E., 1934, Preliminary notice of new manlike apes from India, *Amer. J. Sci.,* 27, 161.

Macintosh, N. W. G., 1967, Recent discoveries of early Australian man, *Ann. Aust. Coll. Dent. Surg.,* 1, 104.

Mann, A. E., 1968, The paleodemography of *Australopithecus,* University Microfilms.

Martyn, J., and Tobias, P. V., 1967, Pleistocene deposits and new fossil localities in Kenya, *Nature,* 215, 476.

Mason, W. A., 1968, Use of space by *Callicebus* groups, IN *Primates,* ed. P. C. Jay, Holt, p. 200.

Mayr, E., 1963, *Animal species and evolution,* Harvard.

Montagna, W., and Ellis, R. A., 1963, New approaches to the study of the skin of primates, IN *Evolutionary and genetic biology of primates,* ed. J. Buettner-Janusch, Academic, p. 179.

Napier, J. R., 1959, Fossil metacarpals from Swartkrans, *Fossil mammals of Africa,* no.17, Brit. Mus. Nat. Hist.

——, **1962a,** The evolution of the hand, *Sci. Amer.,* 207, 56.

——, **1962b,** Fossil hand bones from Olduvai Gorge, *Nature,* 196, 409.

——, **1964,** The evolution of bipedal walking in the hominids, *Arch. Biol.* (Liège), 75, 673.

——, **1967,** The antiquity of human walking, *Sci. Amer.,* 216, 56.

——, **and Napier, P., 1967,** *A handbook of living primates,* Academic.

——, **and Walker, A. C., 1967,** Vertical clinging and leaping: A newly recognized category of locomotor behavior of primates, *Folia Primat.,* 6, 204.

Oakley, K. P., 1961, *Man the tool maker,* 5th ed., Brit. Mus.. Nat. Hist.

——, **1966,** Discovery of part of skull of *Homo erectus* with Buda Industry at Vértesszöllös, Northwest Hungary, *Proc. Geol. Soc. Lond.,* 1630, 31.

Oxnard, C. E., 1969a, Evolution of the human shoulder: Some possible pathways, *Amer. J. Phys. Anthrop.,* 30, 319.

——, **1969b,** Mathematics, shape, and function: A study in primate anatomy, *Amer. Sci.,* 57, 75.

Patterson, Bryan, Behrensmeyer, A. K., and Sill, W. D., 1970, Geology and fauna of a new Pliocene locality in north-western Kenya, *Nature,* 226, 918.

Patterson, B., and Howells, W. W., 1967, Hominid humeral fragment from early Pleistocene of Northwestern Kenya, *Science,* 156, 64.

Pilbeam, D. R., 1968, The earliest hominids, *Nature,* 219, 1335.

——, **1969a,** Tertiary Pongidae of East Africa: Evolutionary relationships and taxonomy, *Peabody Mus. Bull.* (Yale), no. 31.

——, **1969b,** Newly recognized mandible of *Ramapithecus, Nature,* 222, 1093.

——, **1970,** *Gigantopithecus* and the origins of Hominidae, *Nature,* 225, 516.

——, **and Simons, E. L., 1965,** Some problems of hominid taxonomy, *Amer. Sci.,* 53, 237.

Pilgrim, G. E., 1910, Notices of new mammalian genera and species from the Tertiaries of India, *Rec. Geol. Surv. India,* 40, 63.

Radinsky, L., 1967, The oldest primate endocast, *Amer. J. Phys. Anthrop.,* 27, 385.

Read, D. W., and Lestrel, P. E., 1970, Hominid phylogeny and immunology: A critical appraisal, *Science,* 168, 578.

Reynolds, P. C., 1969, Evolution of primate vocal–auditory communication systems, *Amer. Anthrop.,* 70, 300.

Reynolds, V., 1965, Some behavioral comparisons between the chimpanzee and the mountain gorilla in the wild, *Amer. Anthrop.,* 67, 691.

——, **1966,** Open groups in hominid evolution, *Man,* 1, 441.

——, **and Reynolds, F., 1965,** Chimpanzees of the Budongo Forest, IN *Primate behavior,* ed. I. DeVore, Holt, p. 368.

Robinson, J. T., 1953, *Meganthropus,* australopithecines, and hominids, *Amer. J. Phys. Anthrop.*, 11, 1.

———, **1954,** The genera and species of *Australopithecus, Amer. J. Phys. Anthrop.,* 12, 181.

———, **1956,** The dentition of the Australopithecinae, *Transvaal Mus. Mem.,* 9.

———, **1959,** A bone implement from Sterkfontein, *Nature,* 184, 583.

———, **1961,** The australopithecines and their bearing on the origin of man and of stone tool making, *S. Afr. J. Sci.,* 57, 3.

———, **1963,** Adaptive radiation in the australopithecines and the origin of man, IN *African ecology and human evolution,* eds. F. C. Howell and F. Boulière, Aldine, p. 385.

———, **1964,** Some critical phases in the evolution of man, *S. Afr. Arch. Bull.,* 19, 3.

Rowell, T. E., **1966,** Forest living baboons in Uganda, *J. Zool.,* 149, 344.

Sade, D. S., **1965,** Some aspects of parent–offspring and sibling relations in a group of rhesus monkeys, with discussion of grooming, *Amer. J. Phys. Anthrop.,* 23, 1.

———, **1968,** Inhibition of son–mother mating among free-ranging rhesus monkeys, *Science and Psychoanalysis,* 12, 18.

Sarich, V. M., **1968,** The origin of the hominids: An immunological approach, IN *Perspectives on human evolution,* vol. 1, eds. S. L. Washburn and P. C. Jay, Holt, p. 94.

———, and **Wilson, A. C., 1968,** Immunological time scale for hominid evolution, *Science,* 158, 1200.

Schaefer, U., **1959,** Die Stellung der Skelette aus Krapina im Rahmen der Neandertaler der Riss-Würm Interglazials und des Würm-Glazials, *Ber. 6 Tag. Deutsch. Gessel. Anth.,* 209.

Schaller, G. B., **1963,** *The mountain gorilla,* U. of Chicago.

Schultz, A. H., **1966,** Changing views on the nature and interrelations of the higher primates, *Yerkes Newsletter,* 3, 15.

———, **1968,** The recent hominoid primates, IN *Perspectives on human evolution,* vol. 1, eds. S. L. Washburn and P. C. Jay, Holt, p. 122.

Scott, J., **1963,** Factors determining the skull form in primates, *Symp. Zool. Soc. Lond.,* no. 10, 127.

Service, E. R., **1966,** *The hunters,* Prentice-Hall.

Simons, E. L., **1960,** New fossil primates: A review of the past decade, *Amer. Sci.,* 48, 179.

———, **1961a,** Notes on Eocene tarsioids and a revision of some Necrolemurinae, *Bull. Brit. Mus. Nat. Hist.,* 5, 6.

———, **1961b,** The phyletic position of *Ramapithecus, Postilla* (Yale), no. 57.

———, **1962,** A new Eocene primate genus, *Cantius,* and a revision of some allied Eurpean lemuroids, *Bull. Brit. Mus. Nat. Hist.,* 7, 1.

———, **1963,** Some fallacies in the study of hominid phylogeny, *Science,* 141, 879.

———, **1964a,** The early relatives of man, *Sci. Amer.,* 211, 51.

———, **1964b,** On the mandible of *Ramapithecus, Proc. Nat. Acad. Sci.,* 51, 528.

———, **1965,** New fossil apes from Egypt and the initial differentiation of Hominoidea, *Nature,* 205, 135.

———, **1969a,** Miocene monkey (*Prohylobates*) from Northern Egypt, *Nature,* 223, 687.

———, **1969b,** Late Miocene hominid from Fort Ternan, Kenya, *Nature,* 221, 448.

———, and **Chopra, S. R. K., 1969,** *Gigantopithecus* (Pongidae, Hominoidea): A new species from North India, *Postilla* (Yale), no. 138.

———, and **Ettel, P. C., 1970,** *Gigantopithecus, Sci. Amer.,* 222, 76.

———, and **Pilbeam, D. R., 1965,** Preliminary revision of the Dryopithecinae (Pongidae, Anthropoidea), *Folia Primat.,* 3, 81.

———, and **Ettel, P. C., 1969,** Controversial taxonomy of fossil hominids, *Science,* 166, 258.

Solecki, R. S., **1963,** Prehistory in Shanidar Valley, Northern Iraq, *Science,* 139, 179.

Stepien, L., Cordeau, J. P., and Rasmussen, T., **1960** The effect of temporal lobe and hippocampal lesions on auditory and visual recent memory in monkeys, *Brain,* 83, 470.

Straus, W. L., **1963,** The classification of *Oreopithecus.* IN *Classification and human evolution,* ed. S. L. Washburn, Aldine, p. 146.

———, and **Cave, A. J. E., 1957,** Pathology and the posture of neandertal man, *Q. Rev. Biol.,* 32, 348.

Struhsaker, T. T., **1969,** Correlates of ecology and social organization among African cercopithecines, *Folia Primat.,* 11, 80.

Tattersall, I., **1969,** Ecology of North Indian *Ramapithecus, Nature,* 221, 451.

Thoma, A., **1966,** Occipital de l'homme Mindélien de Vertesszöllös, *L'Anthropol.,* 70, 495.

Tobias, P. V., **1960,** The Kanam jaw, *Nature,* 185, 946.

———, **1961,** New evidence and new views on the evolution of man in Africa, *S. Afr. J. Sci.,* 57, 25.

———, **1962,** Early members of the genus

Homo in Africa, IN *Evolution und Hominisation,* ed. G. Kurth, Fischer, p. 191.

———, **1966,** The distinctiveness of *Homo habilis, Nature,* 209, 953.

———, **1967,** *Olduvai Gorge,* vol. 2, Cambridge U. P.

———, **1968,** Cranial capacity in anthropoid apes, *Australopithecus,* and *Homo habilis,* with comment on skewed samples, *S. Afr. J. Sci.,* 64, 81.

———, and **Hughes, A. R., 1969,** The new Witwatersrand University excavation at Sterkfontein, *S. Afr. Arch. Bull.,* 24, 158.

———, and **von Koenigswald, G. H. R., 1964,** A comparison between the Olduvai hominines and those of Java and some implications for hominid phylogeny, *Nature,* 204, 515.

Tuttle, R. H., 1967, Knuckle walking and the evolution of hominoid hands, *Amer. J. Phys. Anthrop.,* 26, 171.

———, **1969,** Knuckle walking and the problem of human origins, *Science,* 166, 953.

Uzzell, Thomas, and Pilbeam, David, 1971, Phyletic divergence dates of hominoid primates: a comparison of fossil and molecular data, *Evolution,* in press.

von Koenigswald, G. H. R., 1940, Neue *Pithecanthropus*-funde 1936–1938, *Wet. Meded. Sienst. Mijub. Ned.-O.-Ind.,* 28, 1.

———, **1957,** Remarks on *Gigantopithecus* and other hominoid remains from Southern China, *Kon. Ned. Akad. Weten.,* 60, 153.

———, **1962,** Das absolute Alter der *Pithecanthropus erectus* Dubois, IN *Evolution und Hominisation,* ed. G. Kurth, Fischer, p. 112.

Walker, A. C., 1967, Ph.D. thesis, London University (unpublished).

———, and **Rose, M. D., 1968,** Fossil hominoid vertebra from the Miocene of Uganda, *Nature,* 217, 980.

Washburn, S. L., 1960, Tools and human evolution, *Sci. Amer.,* 203, 63.

———, **1968,** *The study of human evolution,* Condon Lectures, Eugene, Oregon.

———, and **DeVore, I., 1961a,** The social life of baboons, *Sci. Amer.,* 204, 62.

———, **1961b,** Social behavior of baboons and early man, IN *Social life of early man,* Aldine, p. 91.

———, and **Lancaster, C. S., 1968,** The evolution of hunting, IN *Man the hunter,* eds. R. B. Lee, and I. DeVore, U. of Chicago, p. 293.

Weidenreich, F., 1936, The mandibles of *Sinanthropus pekinensis, Palaeont. Sinica,* ser. D, 1.

———, **1937,** The dentition of *Sinanthropus*

pekinensis, Palaeont. Sinica, ser. D, 7.

———, **1941,** The extremity bones of *Sinanthropus pekinensis, Palaeont. Sinica,* no. 116.

———, **1943,** The skull of *Sinanthropus pekinensis, Palaeont. Sinica,* no. 127.

———, **1945,** Giant early man from Java and South China, *Anthrop. Papers, Amer. Mus. Nat. Hist.,* 40, 1.

———, **1951,** Morphology of Solo man, *Anthrop. Papers, Amer. Mus. Nat. Hist.,* 40, 1.

Weiner, J. S., and Campbell, B. G., 1964, The taxonomic status of the Swanscombe skull, IN *The Swanscombe skull,* ed. C. D. Ovey, *Roy. Anth. Inst. Gr. Brit.*

Wolpoff, M. H., 1968, "Telanthropus" and the single species hypothesis, *Amer. Anthrop.,* 70, 477.

Woo, J. K., 1964, Discovery of the mandible of *Sinanthropus lantianensis* in Shensi Province, China, *Curr. Anthrop.,* 5, 98.

———, **1966,** The skull of Lantian man, *Curr. Anthrop.,* 7, 83.

Zapfe, H., 1958, The skeleton of *Pliopithecus (Epipliopithecus) vindobonensis* Zapfe and Hürzeler, *Amer. J. Phys. Anthrop.,* 16, 441.

Zihlman, A. L., 1967, Human locomotion: A reappraisal of the functional and anatomical evidence, University Microfilms.

Index